Hubble

The mirror on the universe

A FIREFLY BOOK

Publishe d by Firefly Books Ltd. 2007

First printing

Publisher Cataloging-in-Publication Data (U.S.)

Kerrod, David.
    Hubble : the mirror on the universe / Robin Kerrod and Carole Stott.
Rev. ed.
Originally published, 2003, by Robin Kerrod.
[192] p. : col. ill. ; cm.
Includes index.
Summary: Follows the design and construction of the Hubble space telescope, including the political and practical difficulties, and the initial optical fault. Explains the workings of the telescope, and how the masses of data generated from it have provided new insights into the structure and evolution of our universe.
ISBN-13: 978-1-55407-316-0
ISBN-10: 1-55407-316-2
1. Hubble Space Telescope (Spacecraft). 2. Orbiting astronomical observatories. 3. Outer space — Exploration. I. Title.
520.22/2 dc22   QB500.268.K477 2007

Library and Archives Canada Cataloguing in Publication

Kerrod, Robin
Hubble : the mirror on the universe / Robin Kerrod and Carole Stott. —Rev. ed.
Includes index.
ISBN-13: 978-1-55407-316-0
ISBN-10: 1-55407-316-2
1. Hubble Space Telescope (Spacecraft). 2. Space astronomy.
I. Stott, Carole II. Title.
QB500.268.K47 2007   522'.2919   C2007-901492-5

Published in the United States by
Firefly Books (U.S.) Inc.
P.O. Box 1338, Ellicott Station
Buffalo, New York 14205

Published in Canada by
Firefly Books Ltd.
66 Leek Crescent
Richmond Hill, Ontario L4B 1H1

This book was conceived, designed and produced by
Quintet Publishing Limited
6 Blundell Street
London N7 9 BH

Editors: Toria Leitch, Marian Broderick
Designers: Roger Fawcett-Tang, Graham Saville
Illustrator: Richard Burgess
Managing Editor: Diana Steedman
Art Director: Tristan de Lancey
Creative Director: Richard Dewing
Publisher: Gaynor Sermon

Color separation in Singapore by Pica Digital Pte Ltd

Cover images courtesy of NASA and STSci.

Printed in China

Hubble

The mirror on the universe

Robin Kerrod & Carole Stott

FIREFLY BOOKS

Contents

Foreword

Everyone who works on the Hubble Space Telescope program encounters people in their everyday life — the dentist, the auto mechanic, a stranger sitting next to you on an airplane — who asks us what we do for a living. When we say, "I work on the Hubble Space Telescope", invariably their eyes light up and they say, with a tone of excitement, something like, "Wow, that's so cool," or more recently, "Gee, I hope you can save Hubble." I'm often asked what it is about Hubble that causes it to be so enthusiastically embraced by so many people — not just scientists, but people from all walks of life. Many explanations have been suggested, probably all true to some extent.

Here's my take on it: we are privileged to be the first generation of *Homo sapiens* to gain a clear and deep view of the visible universe. And what we see "out there" is staggering in its beauty, awesome in its scale, and shocking in the way it has upended our preconceived notions about how nature works. You don't have to be a scientist to grasp this. Any thinking person who has come in contact with Hubble images and Hubble discoveries seems to find exhilaration in the notion that our place in the grand scheme of things is now better defined than in all of human history to date.

There are some other important factors. The Hubble Space Telescope belongs to all of humanity. It is an international facility that any scientist in any country can competitively propose to use. In the process of peer review, the best ideas win time on the telescope, regardless of where they came from. At the same time, any person on earth who has access to the internet can peruse the entire archive of Hubble data, not to mention the large store of Hubble imagery and other material specifically aimed at a non-professional audience. Any student at any level at any school in the world at any time can gain knowledge and inspiration from Hubble observations. For the many of us who are lifelong fans of science fiction, Hubble is the surrogate starship that transports us across the Universe when there is, as yet, no other way to make the journey. It gives flight to our imagination and creativity.

Finally, we as human beings can take justified pride in the fact that we have created and used Hubble for entirely peaceful purposes in a world that suffers continuous conflict and pain. Hubble is noble. And we made it!

David S. Leckrone
Senior Project Scientist
Hubble Space Telescope
NASA, Goddard Space Flight Center
June 5, 2007

page 5
Cauldron of creation
These swirling clouds of gas and dust in the Tarantula Nebula are vast stellar nurseries, spawning stars by the thousand. This nebula resides in our galactic neighbor the Large Magellanic Cloud.

left
Space walk
Astronauts replace one of the two Hubble Space Telescope's second-generation solar arrays, and a Diode Box Assembly.

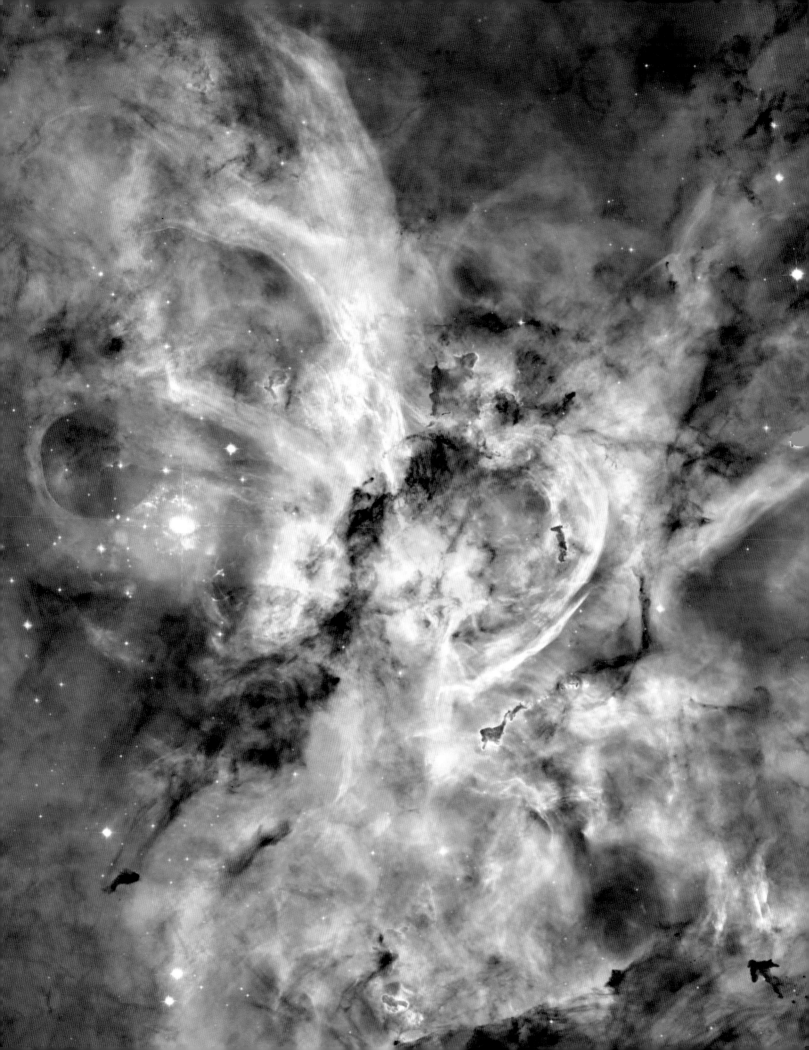

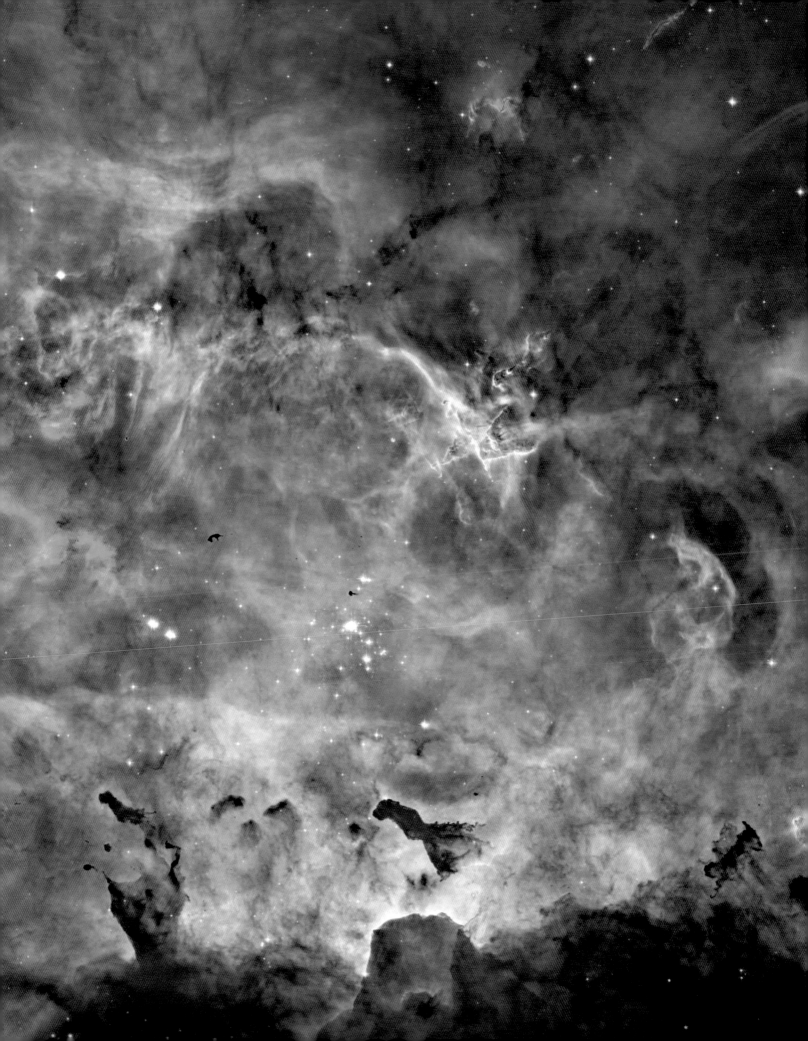

Hubble: the story so far

For almost two decades the Hubble Space Telescope (HST) has enabled us to see and understand our universe more clearly. It has also helped us to place our home and our lives in a wider context, balancing our concept of the vastness and permanence of planet Earth with the realization that we actually live on a rather small planet and that half the planets in our solar system are much bigger. Hubble has allowed us to probe the childhood of our universe and witness the formation of galaxies, investigate the 200 billion stars in our own galaxy, see the extraordinary spectacle of star birth and death, and explore objects in our solar system.

The HST Flight Operations Team is located at the Goddard Space Flight Center in Greenbelt, Maryland. The telescope is observing 24 hours a day, 7 days a week, the only respite being when the instrument is visited by astronauts during the servicing missions. The amount of work done by the Hubble Space Telescope is prodigious: during its first seventeen years of operation (May 1990–May 2007) it has taken nearly 500,000 astronomical images of more than 25,000 celestial bodies. The exposure times of each image varies drastically depending on the faintness of the body, but this production rate is equivalent to an average of 3.4 images per hour. The astronomer whose observation program is being carried out by the telescope has exclusive rights to analyze the image data for a year, but then the data is archived and made available to everyone. So far, nearly 7,000 scientific papers have been published using Hubble data. Every year more than 1,000 observation proposals are submitted by astronomers from all over the world, and around 200 are selected.

Beginnings

The development of the Space Shuttle in the 1970s revolutionized space telescope design, making it possible to use astronauts to aid the delivery of telescopes to orbit and allow mechanisms to be serviced, failed components replaced, and new and more up-to-date instruments mounted in the focal plane. Not only could a telescope start out as a state-of-the-art design, it could remain so throughout its fifteen-year planned lifetime.

The NASA design study for the Hubble Space Telescope commenced in 1973 and the European Space Agency joined the project in 1975. Construction and assembly took nearly a decade, and the whole spacecraft was finished in 1985. Launch was scheduled for 1986, but the Challenger accident on January 28th—where the seven crew members of the STS-51-L mission were killed when the shuttle disintegrated just 73 seconds after take off—and the subsequent Shuttle redesign put back the Kennedy Space Center launch until 24 April 1990. "First light," a momentous time for any telescope, occurred on 20 May 1990. Very quickly it was realized that there was a flaw in the main mirror that introduced spherical aberration and blurred the stellar images. Corrective optics, rather like a huge

page 8-9
Extreme star birth
The seventeenth anniversary of HST's launch was celebrated by the release of this 50 light-years wide image of the interstellar nebulosity near the star Eta Carinae (far left), some 75 light years away. The remnants of the original cloud of gas and dust from which the stars were born is being sculpted by out-flowing stellar winds and the effect of UV radiation pressure. Star-birth started about 3 million years ago and is ongoing in regions compressed by bubbles of hot gas.

left
Venus Express
The launch of Starsem flight ST14 and a Soyuz-Fregat rocket launcher carrying Venus Express. The mission successfully launched from Baikonur launch pad number 6 , Kazakhstan, on November 9 2005.

contact lens, was shipped up to the telescope on the first service mission in December 1993. A new wide field and planetary camera was also installed and sharp images started to be produced. In February 1997 seven astronauts flew to the Hubble Space Telescope and replaced two of the four focal plane instruments. The first replacement, the Near Infrared Camera and Multi-Object Spectrometer (NICMOS) had detectors cooled by evaporating solid nitrogen. Infrared radiation passes through interstellar dust more efficiently than visual radiation and so regions where new stars and planets were being formed could be probed in more detail. The second new instrument, the Space Telescope Imaging Spectrograph, could take detailed spectra of 500 places across a specific astronomical region. These spectra were then used to estimate the chemical composition, temperature, and relative velocity of the regions in question.

Six gyroscopes on board the HST help it maintain precision pointing. These were all replaced on the third servicing mission in mid December 1999. A new central computer and new Fine Guidance Sensor replaced older, worn, and less sensitive versions. March 2002 saw astronauts back again. Another Wide Field and Planetary Camera was introduced.

By doing this the field of view (the area of sky being imaged) was doubled, and the speed at which data could be collected was increased by a factor of ten. As time passed, the telescope simply got better and better.

The Future of Hubble

The final trip to the Space Telescope is planned to take place some time between spring and fall 2008. This visit will ensure that HST works even better than before, at least up to the year 2013. The sensitivity to the ultra-violet radiation is going to be improved by replacing one of the existing instruments with the Cosmic Origins Spectrograph. This will probe the distribution of galaxies and intergalactic gas to see just how the distribution of dark matter affects the large-scale structure of the universe. A more sensitive wide field camera will also be introduced. The Space Telescope Imaging Spectrograph that was new in 1997 stopped working in 2004, and it is hoped that the astronauts will be able to repair it in 2008. The ability to be continually upgraded and improved is one of the great advantages of the Hubble Space Telescope.

What is special about the year 2013? Well, space is an extremely harsh environment and space instrumentation does

above
Astro-ant
No-one knows why the central star of the planetary nebula Menzel 3 — the Ant Nebula — ejects matter into such a strange pattern. It could be a magnetic effect, or there may be two stars at work.

not last forever. By 2013 the Hubble Space Telescope will have been working for 23 years. It will be time for a replacement and this is known as the Next Generation Space Telescope (NGST). 2013 is the year scheduled for the launch of the James Webb Space Telescope (the NGST being named after a former NASA administrator). The JWST is massive and is funded by the USA, Europe, and Canada. Instead of an HST main mirror 2.5 m across the JWST will have a mirror that is 6.5 m in diameter, giving it seven times more radiation collecting area. The JWST mirror is made of eighteen segments that will be gently unfolded and adjusted to shape when in orbit. The JWST will not travel around Earth fifteen times a day as Hubble does; it will orbit around a position in space known as the second Lagrangian point, this being 930,000 miles from Earth, way beyond the Moon's orbit and diametrically opposite to the Sun in the sky. Needless to say, when in orbit, the JWST is completely beyond the reach of manned spacecraft. If anything goes wrong, no shuttle mission could be sent to fix it. The mirror and instrumentation will be cold and shielded from solar and terrestrial radiation by an enormous baffle about the size of a tennis court. The JWST is optimized for infrared observation and will have four science instruments

on board that will concentrate on taking images and spectra. The plan is to probe the primordial early days of galactic evolution and to investigate the mechanisms responsible for star and planetary formation. After the hopefully successful launch and deployment of the JWST in 2013, the HST will most likely lose most of its funding and be "de-orbited" — its small station-keeping rocket systems will drive it into the Earth's upper atmosphere where it will burn up in a startling fireball.

The Hubble Space telescope has been a huge success. There has been drama; corrective optics had to be supplied in the early days. There has been much bravery; traveling in the Space Shuttle, and working in low Earth orbit is not for the fainthearted. There have been breakdowns and instruments have needed to be replaced. But dedication and teamwork has triumphed, and the astronomical productivity has been enviable. When we look back on the history of the telescopic investigation of our universe, the Hubble Space Telescope will always be remembered with pride. This book is a record of Hubble's momentous achievements, and will take you on a journey through time and space to view our wonderful, mysterious, and breathtakingly beautiful universe through the eyes of this extraordinary space telescope.

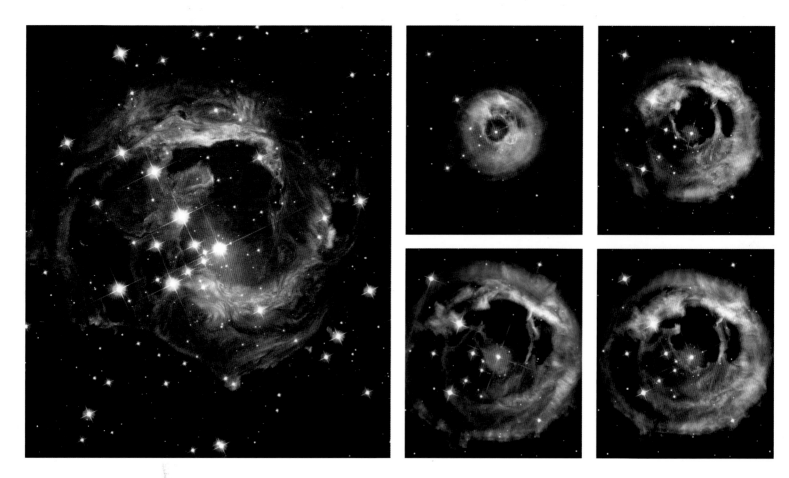

Introduction

Before we look at the universe through the HST's supersensitive eyes, let's set the scene. What, broadly speaking, is this universe of ours like? One thing for sure is that it is vast — unimaginably vast. The Earth, Moon, planets, Sun, and stars are nothing but tiny specks of matter floating in an unfathomable immensity of space — minuscule, insignificant plankton floating in an infinitely deep cosmic ocean.

When we cast our eyes up to the night sky and see a glittering firmament of twinkling stars set in the velvety blackness that is space, we are looking out at one little corner of this ocean, of this universe of ours.

Dominating the sky by night is the silvery Moon, Earth's closest companion in space — its only natural satellite. An airless, heavily cratered ball of rock, it is the only other world that human beings have set foot on — yet.

Next come the ultrabright stars that seem to wander through the heavens. But they are not stars at all: they are planets. What an extraordinary collection of bodies the planets are. They are all quite different from the planet we know best — planet Earth. Closer to the Sun, Mercury, and Venus are oven-hot, while Mars farther out is cold, but might once have been warmer and supported some kind of life. Farther out still are gigantic Jupiter and other gas giants. And farthest out, are the dwarf planets, Plutoand Eris, two small ice worlds.

Dominating the sky by day is the golden orb of the Sun, which brings warmth, light, and life to Earth. It also dominates near space with its powerful gravity, keeping the planets and a host of smaller bodies circling round it. All these bodies make up the Sun's family, or solar system.

The Sun is quite a different body from the planets: it is a huge globe of searing hot gas. It is our local star, just like the myriad other stars in the heavens but very much closer. The other stars, those pinpricks of light in the night sky, are so far away that their light takes years to reach us on Earth (light from the Sun takes just over 8 minutes). Astronomers use the distance light travels in a year (about 6 trillion miles/ 9.5 trillion km) as a measure for expressing distances in space. They call it the light-year. It is the unit we use throughout the book.

The Sun is a very ordinary star, of about average size (nearly 1 million miles/1.6 million km across) and average brightness. There are stars that are very much bigger and brighter, and others that are very much smaller and dimmer.

Since the dawn of astronomy at least five millennia ago, stargazers have used patterns made by the bright stars to guide them across the night sky. These patterns are the constellations. Astronomers use Latin names for the constellations, which refer to figures ancient stargazers thought they could see in the patterns of stars. A few constellations live up to their names (Leo, the Lion; Scorpius, the Scorpion; Cygnus, the Swan), but most don't.

All stars are born in great billowing clouds of gas that occupy the space between the stars. After shining steadily for

above

Echoes of light

Astronomers are extremely lucky that the HST was in orbit at the time when the eruptive variable star V838 Monocerotis was seen to outburst in January 2002. This unusual star, 20,000 light years away, suddenly became 600,000 times more luminous than our Sun. At this specific eruption its surface expanded rapidly but did not escape. In previous eruptions the star has surrounded itself with shells of ejected gas. HST has been observing the star since 2002. The four smaller images show the progression from May 2002 to Ocotober 2004, clockwise, from top left.

perhaps tens of billions of years, stars begin to die. They may exit the universe relatively quietly, as the Sun will eventually, or disappear in a fantastic supernova explosion. The end products of their death throes will be superdense bodies like white dwarfs and neutron stars, or the most awesome objects we know in the universe — black holes.

The stars we see in the night sky may lie many thousands of light-years away, but they are still close neighbors in the universe. They all belong to a great star island in space — a galaxy. The universe is made up of innumerable island galaxies, separated by virtually empty space.

Our galaxy, called the Milky Way or just the Galaxy, probably contains at least 500 billion stars. It measures some 100,000 light-years across and has a spiral shape. Many galaxies are like it, but others are elliptical in shape or have no regular shape at all. We can see just three galaxies in the night sky with the naked eye. They are the Magellanic Clouds in the far southern skies and the Andromeda Galaxy in northern skies.

Some galaxies are extraordinary, pumping out much more energy into the universe than usual, particularly at radio wavelengths. Called active galaxies, they include enigmatic bodies such as quasars and blazars. Black holes seem to be the engines that generate their exceptional power.

Just as stars group together to form great stellar island galaxies, so galaxies themselves group together to form clusters. Our own Galaxy is part of a relatively small group of about 30 galaxies. But we know of clusters containing thousands of galaxies. In their turn, even clusters group together to form superclusters. And on the largest scale, it is strings of superclusters, interspersed with empty voids, that make up the universe.

How can we put such an enormous universe in perspective? With great difficulty — but we can try. Let's suppose we have been able to build an interstellar and intergalactic starship, capable of traveling at the speed of light, and take an incredible journey into space. Setting off from Earth, we would reach the Moon in $1\frac{1}{2}$ seconds and flash pass Venus in $2\frac{1}{2}$ minutes. In less than $8\frac{1}{2}$ minutes we would be leaving the Sun behind, heading for the distant dwarf planet Pluto. We would reach this tiny world in $5\frac{1}{2}$ hours. But it would take several months longer before we escaped completely from the gravitational influence of the Sun and left the solar system. Now traveling in interstellar space, we wouldn't reach even the nearest star (Proxima Centauri) for more than 4 years.

To explore our Galaxy would require flight times measured in tens of thousands of years — 25,000 years to reach the Galaxy's center, twice as long again to reach its edge. To make a visit to our galactic neighbor, the Andromeda Galaxy, we would have a journey time of 2.5 million years. And to reach the farthest objects we can see in the universe, we would have to journey for at least 12 billion years. This is nearly as long as the universe has existed.

1 | Stars in the Firmament

Dark, dusty clouds are the birthplace of stars

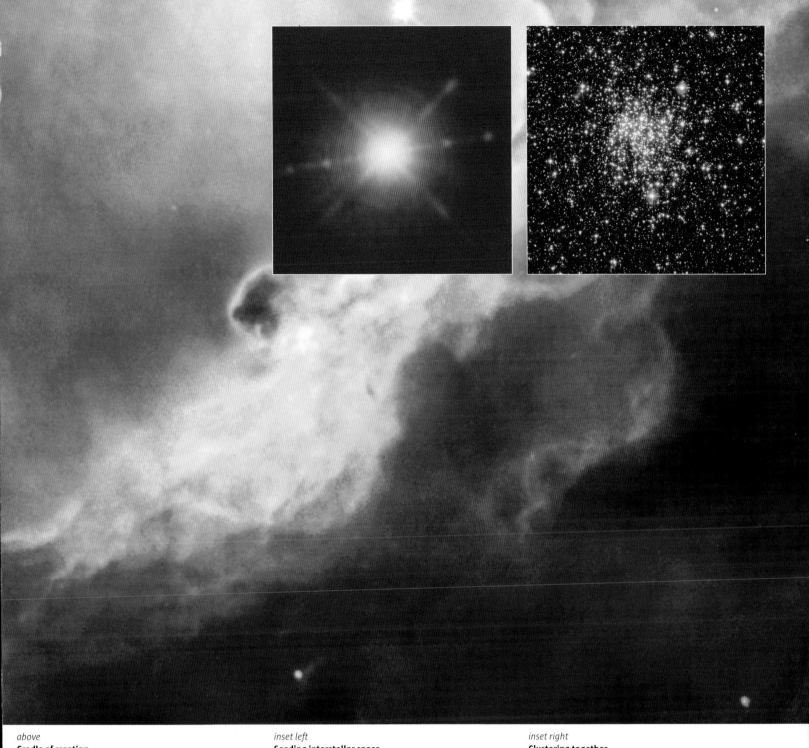

above
Cradle of creation
The Omega Nebula (M17) is a hotbed of new-born stars, wrapped in colorful blankets of glowing gas. The patterns are lit by ultra-violet radiation from young, massive stars just outside the picture. This image was released on 24 April 2003 to commemorate the HST's launch 13 years earlier.

inset left
Seeding interstellar space
Jets of gas stream from an enigmatic object known as He2-90 in Centaurus. It seems to be a close pair of dying stars masquerading as a single youngster — most stars emit jets in their youth. Dying stars add substance to the interstellar medium by shrugging off large amounts of gas and dust.

inset right
Clustering together
A dazzling cluster of stars in one of our galactic neighbors, the Small Magellanic Cloud. These relatively young stars in the cluster (NGC 265) were born from the same cloud of interstellar gas and dust but will drift apart. The image was taken with the HST's Advanced Camera for Surveys.

The Universe of Gas and Dust

In every direction we look in the night sky, we see stars. If we were very patient and very meticulous, we could count around 3,000 stars visible to the naked eye. Looking through binoculars, we would see thousands more stars, and, through a telescope, stars in their billions.

Even a casual glance reveals that stars are not all the same. Some are bright, others dim; most are white, others colored. Brightness and color are two factors that make stars different. But when we investigate stars closely with telescopes and analyze their faint light, we find that they also differ in many other ways—in temperature, mass, speed, magnetism, composition, age and so on. In particular, other stars can differ remarkably from the star we know best—our local neighborhood star, the Sun.

So many kinds of stars. Tiny stellar dwarfs that glimmer feebly, like celestial glowworms; supergiant stars thousands of times bigger and millions of times more brilliant that stand out like cosmic beacons; stars seemingly with a dimmer switch that makes them vary in brightness; newborn stars with the exuberance of youth; ancient stars racing toward celestial Armageddon.

There are so many different species in the stellar zoo. But in studying these disparate bodies, astronomers discover that each kind represents a different stage of stellar evolution. From this, they can piece together the life cycle of a typical star, from its birth, through its youth, into middle age and finally to its death. Astronomers have to do things this way for they can't follow a single star through all its evolutionary stages, which, of course, would take billions of years.

The story of the stars begins not with the stars themselves, but in the vast space that exists among the stars. We tend to think of this space as being empty, but it isn't—quite. Scattered about interstellar space are atoms of hydrogen gas and tiny grains of dust. On average, there are a few dozen hydrogen atoms and the odd speck of dust in a volume of space the size of a soda can. This means that interstellar space is millions of times more empty than the most perfect vacuum that scientists can achieve on Earth.

The tenuous mixture of gas and dust among the stars is known as the interstellar medium, and its typical composition is called the cosmic abundance of the elements. Hydrogen makes up about 75 percent of the interstellar medium, and helium makes up about 23 percent. The remaining two percent is made up of traces of heavier elements, which astronomers rather confusingly call "metals," though this term also encompasses elements such as carbon and oxygen, which scientists consider non-metals.

Carbon and oxygen are among the most abundant of these "metals," carbon being the primary constituent in interstellar dust grains. Nitrogen and iron are relatively common as well. There are also sprinklings of all the other heavier chemical elements—from arsenic to tin, lead to gold.

Where exactly do all of these elements come from? The hydrogen and helium are primordial—they were formed at the very beginning of the universe. All the other elements, however, have been forged in the interior of stars. Ordinary stars such as the Sun produce carbon and oxygen as they enter old age. Supergiant stars produce iron. But all heavier elements are produced in the supernova explosions that mark the death of supergiants.

All these elements are puffed or blasted into the interstellar medium when a star dies and subtly changes its composition.

above
"Cleft into three"
The Trifid Nebula (M20) in Sagittarius is a glorious object through a telescope. It is named for its prominent dark dust lanes, which divide it into three. Radiation from a hot central star within the nebula makes its gas (hydrogen) glow pink.

right (main image)
Stellar nursery
Three huge dark lanes of gas and dust in the Trifid Nebula stand out against clouds of glowing gas. A group of recently formed massive bright stars is seen at the intersection of the lanes. The stars illuminate a dense pillar of gas and dust. At the upper left tip of the pillar, there is a complex filamentary structure (detail lower right). Nearby, a very young star is still surrounded by a ring of gas and dust left over from the star's formation (detail lower left). Near the upper edge of the image a jet of material is being ejected from a very young, low-mass star.

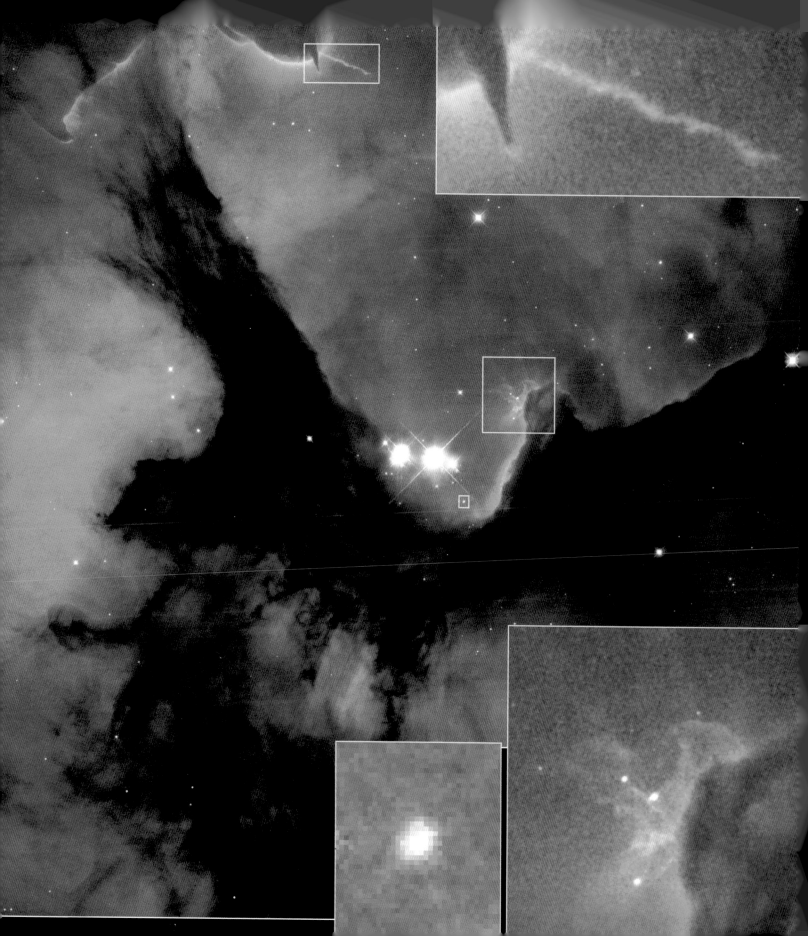

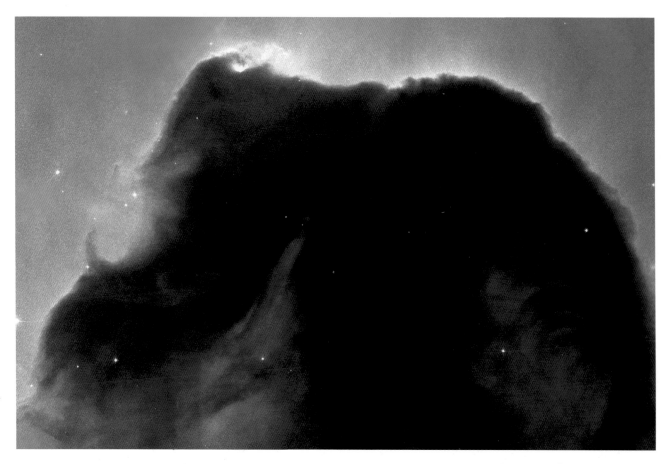

left
Ponderous pillar
Named for its striking shape, the Cone Nebula in Monoceros is a huge pillar of gas and dust. It is lit up by ultraviolet radiation from young, hot stars behind. This was one of the first images taken by the ACS in April 2002 to demonstrate the new instrument's superlative performance.

right
The horse's head
One of the most spectacular of all dark clouds, the Horsehead Nebula in Orion, also known as Barnard 33. With its uncanny resemblance to a horse's head, the cloud is silhouetted against a bright nebula (IC 434) close to Zeta Orionis, the most southerly star in Orion's Belt.

Molecular Clouds

In many regions of interstellar space, the gas and dust becomes relatively dense, though still nowhere near as dense as air. It then forms into vast cloud-like masses. Typically, these clouds are very still and cold, around minus 430 degrees Fahrenheit (–260°C). Under these conditions, the atoms of the elements present join to form molecules, which is why these cool, dense regions are called molecular clouds.

The most common molecule in the clouds is, of course, hydrogen. Scientists have identified many other interstellar molecules, usually by investigations done at infrared, microwave and radio wavelengths. Every molecule has a characteristic wavelength "signature" that identifies it.

Among the other molecules found in molecular clouds are water, ammonia, hydrogen sulfide, formic acid, methanol and glycine. The presence of glycine is particularly interesting because this compound is an organic substance called an amino acid. And amino acids are the building blocks of proteins, which are essential to life as we know it. So the presence of glycine and the other organic chemicals in interstellar space suggests that there may be forms of life elsewhere in the universe.

The Dark Side

Dark molecular clouds lurk throughout the Galaxy. Most merge into the inky blackness of space. But some become visible when they blot out the light of stars or glowing gas behind them. We call these molecular clouds dark nebulas—from the Latin word for clouds.

We can see two huge dark nebulas with the naked eye. One, called the Cygnus Rift, lies in the constellation Cygnus. The other, appropriately named the Coal Sack, lies in the far southern constellation Crux, the famous Southern Cross. Best-known among dark nebulas visible with telescopes is the Horsehead in the constellation Orion. It really does look just like a horse's head and flowing mane.

All Lit Up

Sometimes the clouds of interstellar gas are lit up by nearby stars and become visible as bright nebulas. They are lit in one of two ways: from the outside or the inside.

When there are stars near the Earth side of a nebula, it reflects their light toward us and is called a reflection nebula. When the stars are embedded in the nebula, radiation from them excites the gas molecules, making them emit their own radiation. We then call it an emission nebula. An emission nebula is characteristically red, as this is the wavelength of light emitted by excited hydrogen atoms.

There are many bright nebulas visible in the heavens. These are predominantly emission nebulas, but they also usually feature smaller regions that glow by reflection. Brightest by far and visible to the naked eye is the Orion Nebula (profiled on page 24).

When you look through a telescope, hundreds of bright nebulas swim into view. Many have distinctive shapes that have earned them popular names, such as the Eagle Nebula in Serpens, the Lagoon Nebula in Sagittarius and the North America Nebula in Cygnus.

following pages
Through the Keyhole
Curving filaments of glowing gas and a lacework of silhouetted cold dark clouds make up the Keyhole Nebula. It forms part of the Eta Carina Nebula—Eta Carinae itself lies just beyond the upper right of the image.

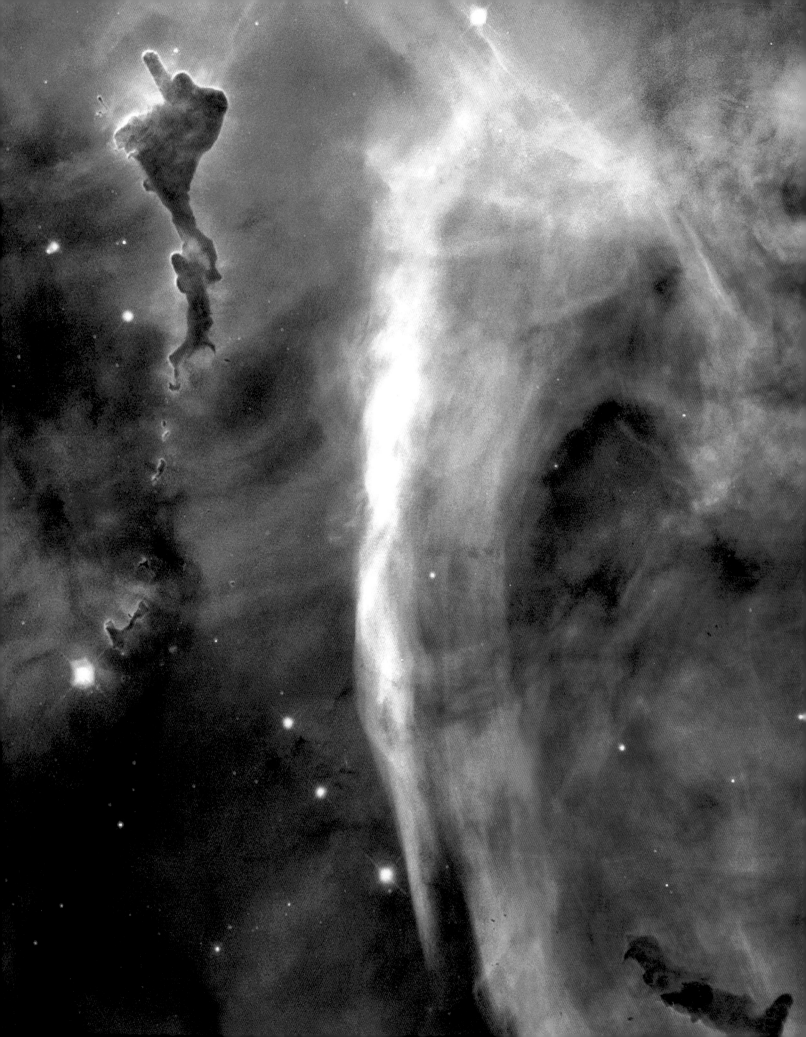

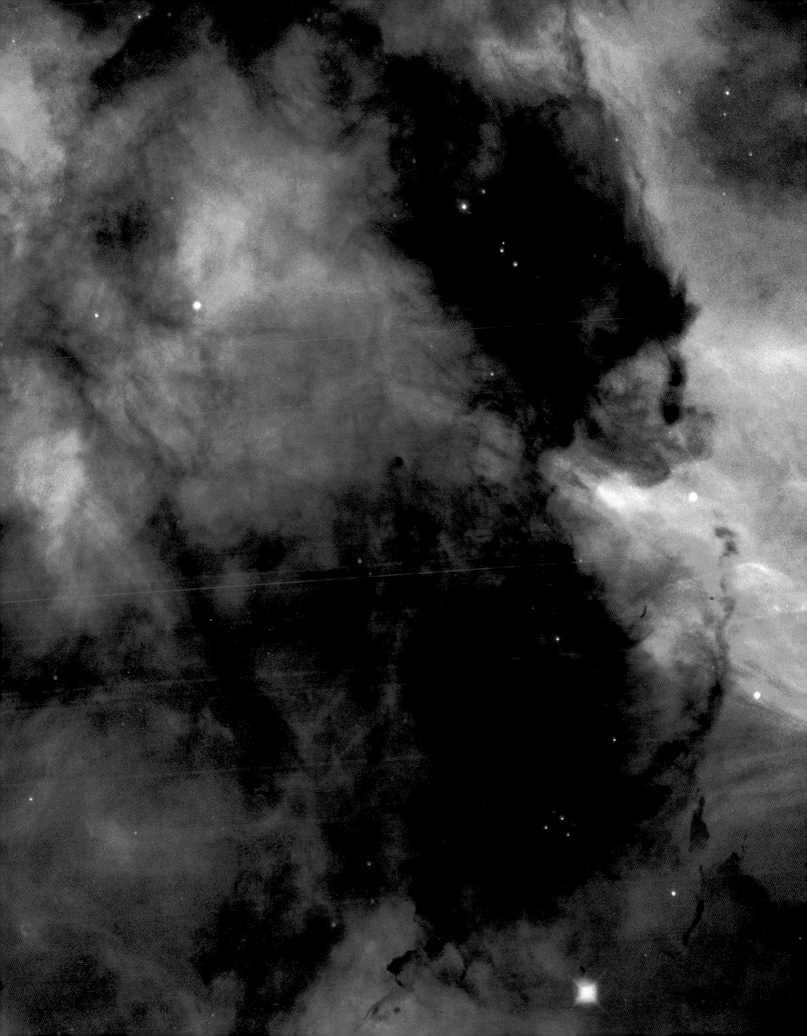

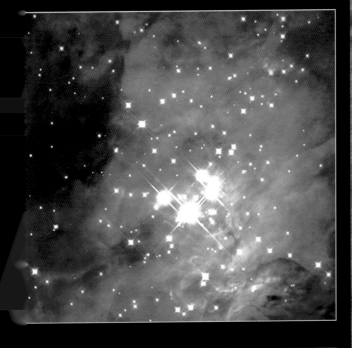

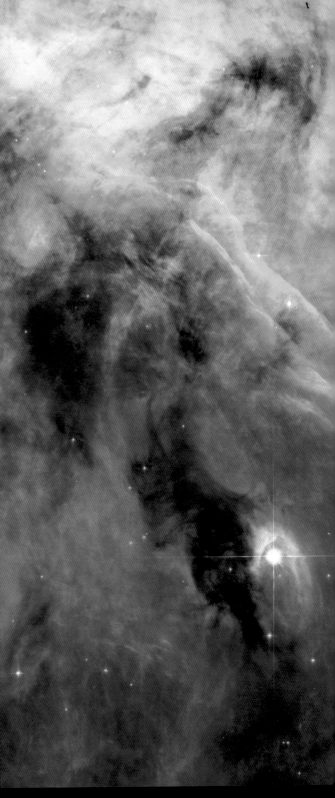

hest constellation in all of the heavens. ... equator, it is visible to stargazers ... . Orion is one of the few constellations ... passing resemblance to the figure it is ... mighty hunter from Greek mythology, ... and in his right a club, raised and

while brilliant white Rigel marks his left knee. Defining Orion's Belt are three slightly less bright stars, and hanging from the belt is his sword.

The sword is marked by a bright, misty patch known as the Orion Nebula, or M42. Binoculars reveal greater detail, and with a small telescope we can make out individual stars. But long-exposure photographs in larger instruments are needed to bring out fully the nebula's incredible beauty.

s that make up Orion are spectacular. The ... ar Betelgeuse marks his right shoulder

Some 1,600 light-years away, M42 is the nearest bright

above
Zooming into the Orion Nebula
This panorama is a part of the Orion Nebula. It is one in a series of detailed views made from the HST's 2006 portrait (left), showing the region just below the center. The Trapezium stars, off the image to the upper left, illuminate the thick swirls and streams of gas

nebula to Earth. It is what astronomers often call a blister nebula — a region at the edge of a dark molecular cloud that has been lit up by the powerful ultraviolet light of nearby embedded stars.

In M42, the star in question is Theta Orionis. Actually, it is a multiple star system, known as the Trapezium because of the arrangement of its four stars. Only about 300,000 years old, these stars are in their infancy, and they are pumping out vast

But this is not the only region in Orion where star formation is taking place. The whole constellation is embedded in vast, billowing clouds of gas and dust. There are two particularly dense areas, known as the southern and northern molecular clouds.

M42 is part of the southern cloud, which nearly merges into the northern cloud around Zeta Orionis, the most southerly star in Orion's Belt. The outstanding feature of the northern cloud is the Horsehead Nebula, which is thrown into dramatic silhouette

A Star Is Born

The cold, dark molecular clouds found in Orion and elsewhere in our Galaxy can stretch for hundreds of light-years. It is in such giant molecular clouds that stars are born. These clouds can remain relatively still and quiescent for hundreds of millions of years. Then something disturbs them, triggering a chain of events that will ultimately lead them to spawn new generations of stars.

Exactly what triggers a disturbance still puzzles astronomers. It could be the shock wave from a nearby supernova explosion, a collision with another giant molecular cloud or a galactic pressure surge. But the end result is the same: Parts of the cloud become so dense that gravity becomes the most dominant force, and they start to collapse.

Barnard and Bok

The collapsing regions in a molecular cloud may contain enough gas and dust to create hundreds of stars. Such regions are known as Barnard Objects, after the US astronomer Edward Barnard. Typically tens of light-years across, they are the kind of dark nebulas that we often see obscuring the light of background stars. We see other, smaller collapsing regions as little black bubbles against a background of stars or a bright nebula. They are called Bok Globules, after the Dutch astronomer Bart Bok. Only one or two light years across, these contain enough mass to make about 20 to 200 sunlike stars.

As both Barnard Objects and Bok Globules collapse, clumps of matter within them collapse as well. So the original cloud fragments successively into smaller and smaller clumps. And it is from the smallest clumps that individual stars form.

DSS ACS

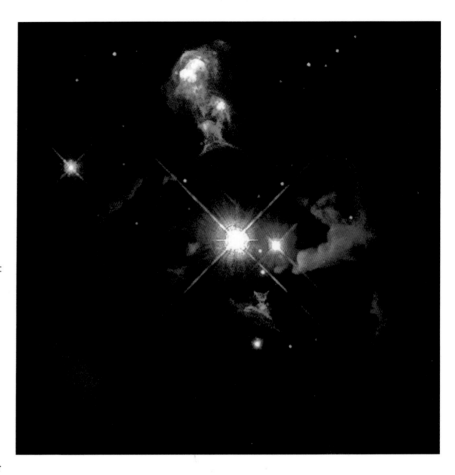

Gravity Rules

In the center (core) of a collapsing clump, concentrations of matter build up. The core is a star in embryo — a protostar. The denser the protostar becomes, the stronger is its gravity and the more it attracts surrounding matter. As gravity tugs at it, matter speeds up; and gravitational energy converts into kinetic energy — the energy of motion.

When this moving matter suddenly reaches the core, it collides with the matter already there and stops. Its energy immediately converts into another form — heat. So, as the protostar accumulates more and more material, its temperature rises.

In a Spin

The protostar is not stationary. The original giant molecular cloud rotates, and it imparts this rotation to the collapsing clump and the protostar within. The spinning mass then flattens out. Over time a thick disk, or ring, of gas and dust builds up around the protostar, while its temperature rises rapidly. It is already giving off radiation, mainly at infrared wavelengths.

As more and more matter rains down, temperatures inside the protostar soar to millions of degrees, and pressures reach millions of atmospheres. Under these exotic conditions, the nuclei (centers) of hydrogen atoms are forced together, and they fuse (join up). Nuclear fusion produces fantastic amounts of energy (see page 118). The protostar springs to life and begins pouring out light, heat and other radiation into the universe. A new star is born.

above
Colorful jets
At the center of this HST image of Herbig-Haro object 32 (HH 32) is a young star blasting jets of matter into space in polar jets. One jet (at the top) is seen plowing into interstellar gas and making it glow in the light of hydrogen atoms (green) and sulfur ions (blue). The jet streaming in the opposite direction is mostly obscured by dust.

left
Bok globule
Dense dark knots of gas and dust called Bok globules are absorbing light in the center of the star forming nebula NGC 281. They are silhouetted against the nebula's pink hydrogen gas. Once they have accumulated enough gas and dust, Bok globules can create stars. But some dissipate before they reach that stage.

Enigmatic "waterfall"
Both polar jets from
a protostar can be
seen slamming into
interstellar gas in
this VLT (Very Large
Telescope) image of
HH 34. The nature of
the prominent waterfall-
like stream of light is
a mystery.

right
Double bubble
A cocoon of gas and
dust surrounds a small
cluster of young hot
stars in the Large
Megallanic Cloud.
This "double bubble"
lies inside the larger
nebula, DEM L106, one
of many star-forming
regions in the galaxy.

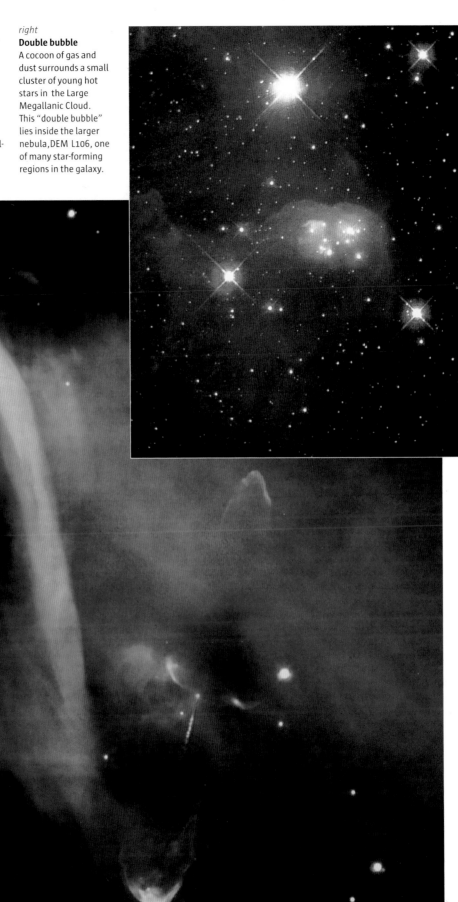

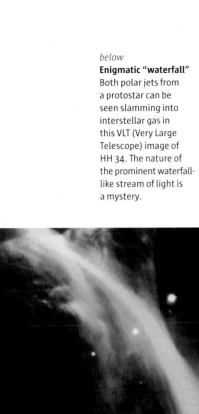

Reaching Equilibrium

Once the nuclear furnace of a star fires up, radiation floods out of it. Until this point, the star has been contracting under gravity. But now radiation pressure starts pushing in the other direction, causing the star to expand. It continues to expand until it reaches an equilibrium point, at which the outward push of the radiation balances the inward pull of gravity. At that point, the newborn star reaches the steady state it will assume for many millions — or even billions — of years. It becomes what astronomers call a main sequence star (see page 35).

In reality, however, things are not that simple. It takes a star quite a while to reach the equilibrium point, typically about 10 million years. When it first starts to expand, battling against gravity, it expands too far and blows material off into space. Then gravity takes charge for a while, forcing the star to contract again. Soon the cycle starts all over again — expansion, mass loss, contraction; expansion, mass loss, contraction. Slowly, the amplitude of these oscillations subsides, until the star finally achieves a state of equilibrium.

The cycle of expansion and contraction causes the star's output of light to fluctuate erratically. This period of a star's life is called the T Tauri phase, after the first star of its kind that was discovered.

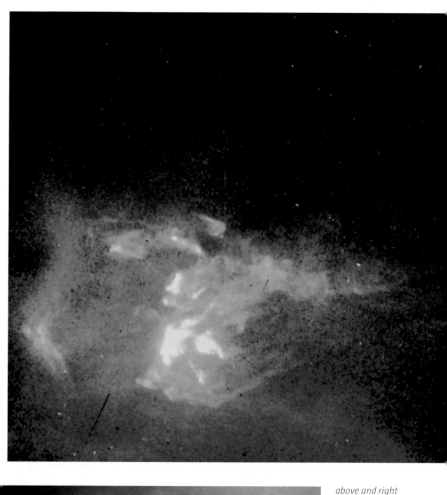

above and right
Jet exhausts
Jets are common "exhaust" products of star formation, revealed when they ram into surrounding gas and dust. In these images, the central stars are hidden within masses of infalling material. Typically, the twin jets from a fledgling star span a region of space about two light-years across.

left
Making waves
Gas streaming from the very young star LL Ori collides violently with the tenuous interstellar medium, creating a "bow shock" around it.

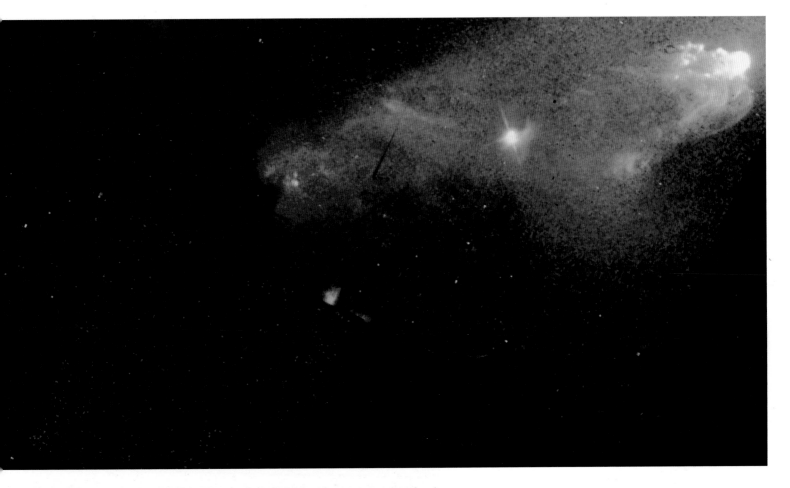

Jet-propelled

In these early days of a star's life, when its nuclear furnace is stoking up, it is still surrounded by a veil of gas and dust and a thick disk of material. It also gives off streams of particles as a kind of stellar "wind" that starts to blow away the surrounding material. This can't happen around the equatorial regions because of the surrounding disk. But it can and does happen at the poles, where matter emerges as powerful jets. The jets stream out into space and collide with interstellar gases, creating "knots" of glowing gas known as Herbig-Haro objects. The jet phase of a star's life is fleeting, lasting only a few thousand years.

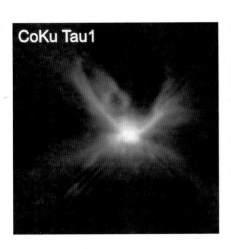

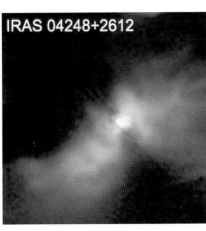

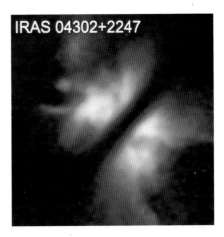

left
Planets in the making
Dark, dusty disks around embryonic stars show up in these infrared images. When the stars begin shining steadily, the disks will probably remain, providing raw materials for future planetary systems.

PROFILING:
The Pillars of Creation

Serpens (the Serpent) is an oddity among the constellations because it is split in two: Serpens Caput—the serpent's head—and Serpens Cauda—the serpent's tail. In between is the constellation Ophiuchus, the Serpent-Bearer.

Ophiuchus, like Serpens itself, is not an easy constellation to identify because it has no particularly bright stars. One of its claims to fame is what it's not—it's not considered by astrologers to be a constellation of the zodiac. And it should be, because the Sun spends longer passing through Ophiuchus than it does passing through Scorpius, which is a zodiac constellation. The omission of Ophiuchus as a star sign (or zodiac constellation) is just one serious weakness in the case for astrology.

But, back to Serpens. Serpens Caput has a fine globular cluster, M15. Serpens Cauda has M16, which is a bright nebula vaguely shaped like a bird with outspread wings. Known as the Eagle Nebula, M16 was the subject of one of the most dramatic images the Hubble Space Telescope (HST) has ever taken.

The HST science team called the image "The pillars of creation." It shows dark columns of gas in which stars are being born. The columns, or pillars, are etched and silhouetted by the light of young, hot, massive stars beyond. The pillar on the left is about one light-year long.

The finger-like protrusions at the top of the pillars are dense regions that probably contain newborn stars or protostars. Termed EGGs (evaporating gaseous globules), they have been revealed because intense ultraviolet radiation from hidden massive stars has blown less dense gas away. Eventually the radiation will blow away the gas in the EGGs as well, revealing the star inside for the first time.

above
Sculpted pillar
This detail of an Eagle Nebula pillar was released by the HST team in April 2005. Light from nearby bright, hot, young stars is sculpting the cloud of gas and dust into intricate forms and causing the gas to glow.

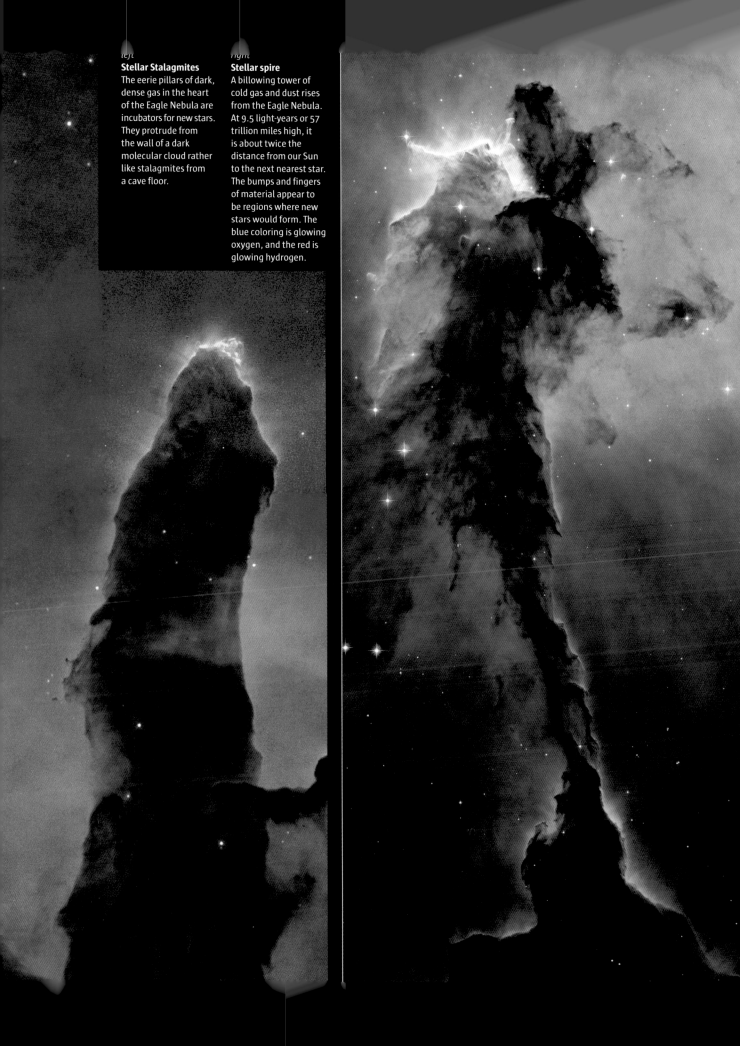

left
Stellar Stalagmites
The eerie pillars of dark, dense gas in the heart of the Eagle Nebula are incubators for new stars. They protrude from the wall of a dark molecular cloud rather like stalagmites from a cave floor.

right
Stellar spire
A billowing tower of cold gas and dust rises from the Eagle Nebula. At 9.5 light-years or 57 trillion miles high, it is about twice the distance from our Sun to the next nearest star. The bumps and fingers of material appear to be regions where new stars would form. The blue coloring is glowing oxygen, and the red is glowing hydrogen.

What Stars Are Like

When a molecular cloud collapses, the stars that form can vary widely in mass. A star's mass determines how long it will shine. Many stars have a mass similar to that of the Sun, which is a relatively small star. But even more stars have a smaller mass—some with less than a tenth of that of the Sun. Stars with much smaller mass don't exist because they don't accumulate enough mass to generate the heat and pressure necessary to trigger nuclear fusion. These "failed stars" glow a dull brown and we call them brown dwarfs.

A collapsing cloud tends to produce relatively few really big stars. But they can be huge, with up to 50 times the mass of the Sun and 20 times its diameter. Bigger stars don't exist because they generate so much energy at the protostar stage that they blow themselves apart.

Mass and Temperature

The mass of a main sequence star determines its temperature. The smallest dwarf stars "burn" their hydrogen fuel relatively slowly and have a low surface temperature, maybe below 5,500 degrees Fahrenheit (3,000°C). Somewhat larger stars like the Sun are about twice as hot, while the most massive giant stars have temperatures beyond 55,000 degrees Fahrenheit (30,000°C).

Allied to a star's temperature is its brightness. Here we are talking about a star's true, or absolute, brightness, not its apparent brightness as we perceive it from Earth. As you might expect, the smallest stars are the dimmest—they may be only 1/100,000,000th as bright as the Sun. And the biggest stars are the brightest—up to 100,000 times brighter than the Sun. This brightness is the star's luminosity, which is compared to that of the Sun.

Temperature and Color

The temperature of a star also determines its color. Think of an iron plunged into a fire: As its temperature rises, it gradually changes color from dull to bright red, then orange, yellow and finally brilliant white. And it is a similar story with stars.

Small, cool stars glow red; we call them red dwarfs, or sub-dwarfs if they are very small. Larger, hotter stars like the Sun glow yellow; we call them yellow dwarfs. The biggest, hottest stars glow blue-white; we call them blue giants.

The Informative Spectrum

When we see a rainbow in the sky, we are looking at a range of different colors, from violet to red. Mixed together, these colors make up white sunlight. The spread of color, or spectrum, is a kind of fanning out of the different wavelengths in light because we perceive different wavelengths as different colors. Violet light has the shortest wavelength, red the longest. Scientists create a spectrum artificially by passing sunlight through an instrument called a spectroscope or spectrograph. The technique is called spectroscopy. The light from all stars can be split into a spectrum in a similar way.

Spectroscopy is one of the most powerful tools in astronomy, since a star's spectrum holds the key to its identity. When you closely examine the rainbow-like spectrum of any star (including the Sun), you notice a series of dark lines. These are called absorption lines, because they are produced when certain wavelengths of starlight are absorbed as they pass through the star's atmosphere. Where these lines appear in the spectrum depends on what elements are present in the atmosphere.

Every element absorbs different wavelengths and gives rise to a characteristic set of spectral lines. So by studying the spectrum, astronomers can work out the composition of a star and much more besides. They can also estimate the star's temperature, density and magnetism; tell how fast it is spinning; and determine whether it is traveling toward or away from us.

Be a Fine Guy

All stars produce a spectrum with dark absorption lines, but each kind of star displays a slightly different and characteristic spectrum, depending on its color (and thus its temperature).

By this means, astronomers divide stars into 10 main groups, or spectral classes, designated O, B, A, F, G, K, M, R, N and S. (Difficult to remember? Try the once-heard, never-forgotten mnemonic "Oh, Be a Fine Guy, Kiss Me Right Now, Sweetie!") O and B stars are the hottest; S stars are the coolest. The cooler the star, the more lines appear in its spectrum.

left
Sirius and companion
The brightest star in the night-time sky, close inspection of Sirius shows it is made of two stars. Sirius A, in the center of this HST image is over-exposed so that its tiny companion, Sirius B (dot at lower left), is visible. Hubble's Space Telescope Imaging Spectrograph has also been used to isolate the light from Sirius B for analysis.

right
Star factory
The nearby dwarf galaxy NGC 1569 is a hotbed of vigorous star birth activity. It contains two massive star clusters as well as a number of smaller ones. It is believed the majority of the clusters were produced in an energetic starburst that lasted 20 million years and only ended a few million years ago.

The H-R Diagram

With such a variety of different stars—bright, dim; hot, cool; blue, red; large, small—how is it possible to classify them? To answer this question, we must go back to two of the leading astronomers from the early part of last century: Ejner Hertzsprung of Denmark and Henry Norris Russell of the United States.

Working independently, they discovered that there is a significant relationship between the true brightness (absolute magnitude) of a star and its spectral class/temperature. When they plotted true brightness against spectral class for a variety of stars, they produced a very interesting graph, which is now called the Hertzsprung-Russell (H-R) diagram.

On the Main Sequence

On the H-R diagram, stars are not scattered about randomly, as you might expect. Instead, they fall into a number of fairly distinct groups. The majority lie within an elongated, S-shaped, diagonal band called the main sequence.

Dim, cool red dwarfs, such as Barnard's Star and Proxima Centauri, are found on the lower right of the main sequence. Bright, hot blue giants, such as Spica and Regulus, are found on the upper left. The Sun, a yellow dwarf, is nearly halfway up the main sequence. In effect, the position of a star on the main sequence depends on its mass. The more massive a star, the hotter it is, because it burns more fuel (hydrogen), faster.

Cool Giants and Hot Dwarfs

Stars are also found in three other main regions on the H-R diagram. Giant stars like Aldebaran and Arcturus are found above the main sequence; they are bright but cool. Supergiants such as Canopus and Antares lie near the top of the diagram; they may be hot and blue or cool and red. The very hot but very tiny bodies we call white dwarfs are at the foot of the diagram.

right
Sparkling jewels
Fourteen massive stars are on the verge of exploding as supernovas in this jewellery box full of sparkling stars. Each is a red supergiant about 20 times as massive as the Sun. They are inside the bluish cluster in the center of the image. A closeup is seen in the inset photo.

Life Cycle

The H-R diagram reflects a frozen moment in the lives of stars we can see in the heavens today. If in a few billion years time we were to plot an H-R diagram again, it would look completely different.

Stars remain on the main sequence for most of their productive lives — that is, when they are burning hydrogen in their cores. But not all stars are the same. Red and yellow dwarf stars remain on the main sequence for billions of years, but hot blue giants pay only a fleeting visit, measured in a few million years.

When stars have used up all their hydrogen, they begin to die and leave the main sequence. The biggest ones become supergiants, the smallest, red giants. Eventually, supergiants blow themselves to bits and disappear from the diagram; some even disappear from the visible universe (as black holes). Red giants shrink and become very hot, turning into white dwarfs. We follow the often spectacular death throes of stars in the next chapter.

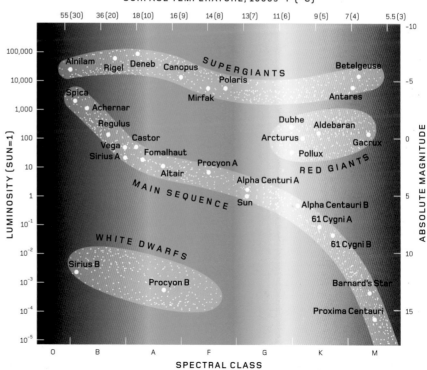

above
Dim dwarfs
Within the ancient globular star cluster NGC 6397 (main image), the HST's Advanced Camera for Surveys has identified the faintest red dwarfs (circled in the lower right image) and the dimmest white dwarfs (circled in the upper right image).

below
The H-R Diagram
A selection of familiar stars plotted on an H-R diagram. They tend to fall into four main groups. Stars in the prime of life ride the main sequence. Giant and supergiant stars lie above this diagonal band, white dwarfs below.

SURFACE TEMPERATURE, 1000s °F (°C)

The H-R diagram plots LUMINOSITY (SUN=1) against SPECTRAL CLASS (O, B, A, F, G, K, M) and ABSOLUTE MAGNITUDE. Stars shown include: Alnilam, Rigel, Deneb, Canopus, Polaris, Betelgeuse (SUPERGIANTS); Spica, Achernar, Regulus, Mirfak, Antares, Dubhe, Aldebaran, Castor, Arcturus, Gacrux, Vega, Pollux, Fomalhaut, Sirius A, Procyon A, Altair, Alpha Centuri A, Sun (MAIN SEQUENCE); RED GIANTS; Alpha Centauri B, 61 Cygni A, 61 Cygni B, Barnard's Star, Proxima Centauri; Sirius B, Procyon B (WHITE DWARFS).

Clustering Together

Stars are born in the giant molecular clouds that exist between the stars, but they are rarely born one at a time. Usually, many clumps of matter condense within a cloud simultaneously, and each one may develop into a star. And if two or more stars are born close together, they may eventually travel as companions through space.

The Sun, of course, has no stellar companion and moves alone in space — except for its family of planets and other bodies. Roughly a third of stars lead solitary lives, but the rest travel with one or more "sibling" stars.

Binaries

Nearly half of all stars are double stars called binaries. In these systems, the two component stars orbit one another in a perpetual celestial waltz. Sometimes we can separate the components — see them separately — in a telescope. But often the components are so close together that we can identify them only by the shifts in their spectral lines as they move to and fro. In this case they are called spectroscopic binaries.

Binaries are not to be confused with optical doubles. These are pairs of stars that are actually far apart, but appear close together only because they happen to lie in the same direction in space as viewed from Earth.

The Winking Demon

A particularly interesting kind of binary is called an eclipsing binary. In this system, the two component stars circle round each other in our line of sight, and each star periodically passes in front of, or eclipses, the other. The overall brightness of the system dips briefly with each eclipse, before regaining its former brilliance when the eclipse is over. Eclipsing binaries are a common kind of variable star — one that varies in brightness. The star Algol ("the winking demon") in Perseus is

a classic example. Other kinds of variable stars, such as Cepheids, vary in brightness because of processes going on inside them (see page 97)

Open Clusters

Sometimes hundreds of stars are born close together and become companions — at least for a while. There are many groups of stars like this in the heavens, called open clusters because the individual stars within them lie relatively far apart.

We can see several open clusters with the naked eye. The easiest to spot is the Pleiades in the constellation Taurus. It is also called the Seven Sisters, because keen-sighted people may be able to see its seven brightest stars. There is also another open cluster in Taurus, called the Hyades. It surrounds (but is not part of) the noticeably orange star Aldebaran, which marks the eye of the bull.

Young Upstarts

Most open clusters are made up of young, hot stars. The Pleiades stars were born about 75 million years ago; those in the Jewel Box in Crux, only about seven million years ago. These clusters are the new kids on the block as far as the universe is concerned, but they will not always stay together. Their gravitational ties are tenuous, and over time they will disperse and gain their independence.

left
Family portrait
This infrared image shows a stellar family of a bright, massive mother star, surrounded by six fainter infants. Violent stellar winds released during the formative years of the mother star triggered star formation in the surrounding gas clouds.

above
Heavyweight stars
Two massive stars, known as Pismis 24-1, orbit one another. They and a third companion, here hidden in the glare of the other two, are estimated to have a combined mass of 200 times that of the Sun. They are amongst the heaviest stars known.

right
Ripples in Tarantula
The star cluster in the image, Hodge 301, is located at the edge of the Tarantula Nebula in the Large Magellanic Cloud. It contains a mix of brilliant, massive stars, including several red supergiants that must be close to exploding as supernovas. Other stars in the cluster have already exploded, sending shock waves rippling through the nebula.

Globulars

In the constellation Hercules is an object designated M13. It is just visible to the naked eye, and clearly seen in binoculars. But a powerful telescope is needed to show it in detail. It proves to be a huge mass of stars packed tightly together into a globe shape and numbered in their hundreds of thousands.

M13 is a concentrated group of stars called a globular cluster. It is not unique. More than 100 globular clusters have been found in our Galaxy, and many more exist in others. The two most spectacular globular clusters in our Galaxy are found in far southern skies, and both are easily seen with the naked eye. Looking rather like hazy stars, they are so identified, as Omega Centauri and 47 Tucanae.

Omega Centauri, in the constellation Centaurus, is thought to contain at least a million stars. It is the biggest globular we know, measuring 180 light-years across. 47 Tucanae, in Tucana, may well have as many stars; it is slightly smaller but has a more concentrated central region.

In the Center

Open clusters like the Pleiades are made up of young, hot stars and are found in the spiral arms of our Galaxy (see page 36). Some of these clusters lie quite close to us.

Globular clusters, however, are made up of old stars and are found toward the center of the Galaxy. They do not take part in the Galaxy's general rotation, but pursue independent orbits around the galactic center. This takes them well above and below the plane of our Galaxy into the spherical region known as the halo.

Whereas the young and hot stars in an open cluster usually appear blue, globular cluster stars appear yellow because they are older and cooler. Astronomers reckon that globulars are typically about 10 billion years old, so they must have formed during or shortly after the formation of the Galaxy itself.

Studying these ancient stellar bundles helps astronomers to chart the early history of our Galaxy. Data collected by the HST plays a particularly crucial role in this study. Individual stars in the very heart of the globulars are resolved (viewed separately). Additionally, the HST has spotted globulars in other galaxies, some of which have proved to be much younger than those in our own.

left
Blue stragglers
This dazzling image shows M80, a globular cluster in Scorpius. It lies nearly twice as far away as 47 Tucanae and contains fewer stars. But it contains a large population of "blue stragglers," which appear to be younger, bluer and more massive than the other stars. It is thought that these bodies have formed from the merger of two colliding stars.

above
The Toucan's gem
A ground-based image of one of the astronomical highlights of the Southern Hemisphere, the globular cluster 47 Tucanae. Containing as many as a million stars packed tightly together, it lies about 15,000 light-years away.

left
In the center
This HST view shows about 35,000 stars in the central region of 47 Tucanae. Most of the stars in the cluster formed about 10 billion years ago.

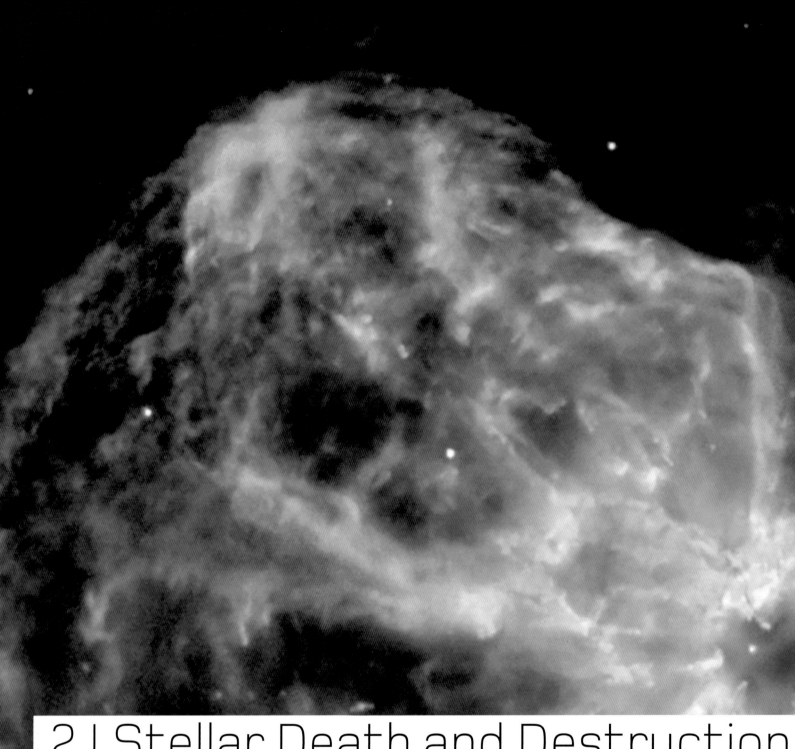

2 | Stellar Death and Destruction

Dying stars exit the universe spectacularly

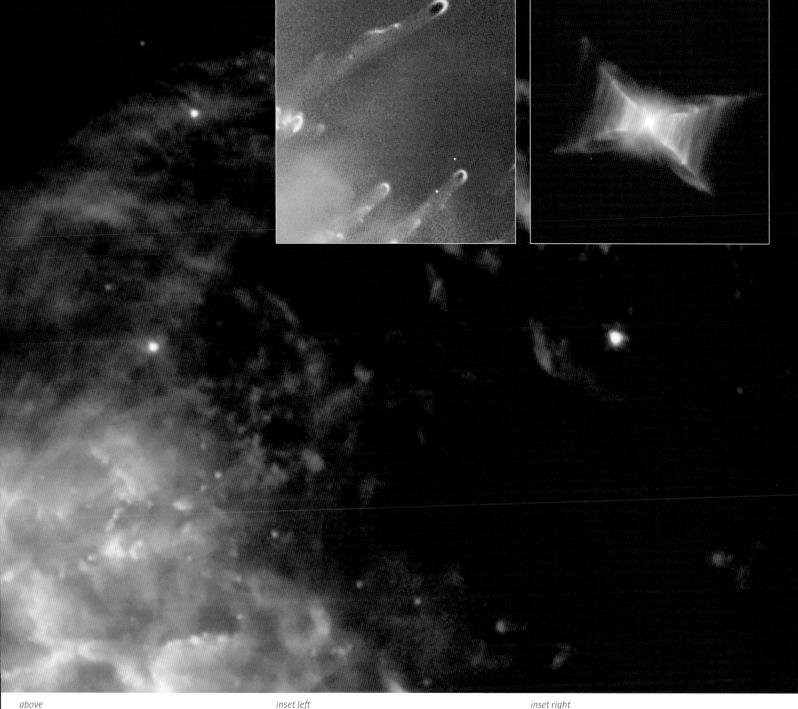

above
Demise of a star
The HST's Wide Field Planetary Camera 2 took this image of a dying star on 6 February 2007. The star is ending its life by casting off its outer layers of gas. The remains of the star, now a white dwarf, is the dot in the center of the colorful gas clouds known as NGC 2440.

inset left
Cometary knots
Tadpole-shaped blobs stream from the inner edge of the Helix Nebula in Aquarius. They are called cometary knots, though they have nothing to do with comets. The Helix Nebula is a close planetary nebula, with gas puffed out by a dying star.

inset right
Red Rectangle
The HST has revealed that the dying star HD 44179, commonly called the Red Rectangle because of its overall shape seen through ground-based telescopes, is not rectangular but has an X-shaped structure thought to be the result of outflows of gas and dust from the central star.

The Beginning of the End

Stars form out of giant clouds of interstellar gas and dust. Then, with their nuclear furnaces stoked up, they join the multitude of other stars on the main sequence. The Sun has spent around 4.6 billion years on the main sequence shining steadily, pouring out light, heat and other radiation into space. So, will it carry on shining like it does today forever? The answer is: No.

Like living things on Earth, the Sun and all the other stars have a natural life-span. Whereas the life-spans of even the oldest living things, such as California's majestic redwoods, are measured in only a few thousand years, those of stars can be measured in tens of billions of years.

The exact life-span of a star depends primarily on its mass. The more massive the star, the shorter is its life. The Sun is a relatively low-mass star, a type known as a yellow dwarf, which has a relatively long life-span of about 9 to 10 billion years. This means that the Sun is now comfortably in middle age. It should carry on shining steadily like it does today for roughly another five billion years. But then, inexorably, it will begin to die.

The Sun and other similar dwarf stars die relatively quietly—we might say, with a whimper. But for stars with much greater mass it is a different story. They depart the celestial scene literally with a bang—the biggest bang in the universe.

A star on the main sequence produces energy by transforming hydrogen into helium in its core. A star like the Sun uses around 600 million tons of hydrogen every second, but it is so massive that it has enough to last for more than 10 billion years.

Eventually, though, the hydrogen in the core of the star runs out, and only helium remains. Nuclear fusion ceases. This heralds an abrupt change in the star's life. It is the beginning of the end—the star is dying.

With no radiation emanating from the star's core, gravity once again becomes the dominant force. The core begins to collapse. The potential energy released by this collapse does two things: it increases the temperature and pressure in the core, and it heats up the outer atmosphere of the star.

Cool Giant

The atmosphere expands greatly, making the star swell up to 30, 50 or perhaps even 100 times its original diameter. Because the star now emits radiation from a much greater surface area, its surface temperature plummets by about half, to around only 5,500 degrees Fahrenheit (3,000°C). This lower temperature makes the star produce a redder light. It becomes a red giant and leaves the main sequence behind (see page 35). When the Sun becomes a red giant in around five billion years time, it will expand 30 times or more and become 1,000 times brighter. It will probably swallow up the planet Mercury and maybe even Venus. The Earth will lose its atmosphere, and the oceans will boil away. Life will be eradicated, and our now-verdant home will become a barren cinder of a planet.

left
Red Sun
Sometimes at sunset, white sunlight turns orange and red as it passes through the lower, dustiest part of the atmosphere. In billions of years, the Sun itself will turn red as it swells up to become a giant star.

below
The generation game
In the giant nebula NGC 3603, the HST has captured stars at different stages in their life-cycles. At upper right are dark clouds that are Bok globules, marking an early stage of star formation. In the center is a cluster of youthful hot stars. Above it is an aging blue supergiant on the brink of destroying itself in a supernova explosion.

left
Dying Egg
Hidden inside this cocoon of dusty gas (the Egg Nebula) is a red giant star. It has been ejecting waves of material for only a few hundred years as it periodically pulsates.

below
In the infrared
Peering inside the Egg Nebula in infrared light, the HST reveals that the dying giant ejects material along preferred axes, in four broad jets.

Spotting Giants

Some of our most familiar stars are red giants, which appear in the sky with a noticeable orange-red hue. Arcturus, in the constellation Boötes (the Herdsman), is a red giant. It is the most brilliant star in the Northern Hemisphere and the fourth-brightest in all of the heavens. You can easily find it by following the curve in the handle of the Big Dipper, the most recognizable part of the constellation Ursa Major (the Great Bear). Capella in Auriga (the Charioteer) and Aldebaran in Taurus (the Bull) are two other northern giants. Aldebaran marks the baleful red eye of the bull, which appears to be charging the mighty hunter Orion in the adjacent constellation.

In the Southern Hemisphere, Gacrux, one of the four bright stars that make up Crux (the Southern Cross), is a red giant. Its warm orange color contrasts with the other three stars in the cross, which are white.

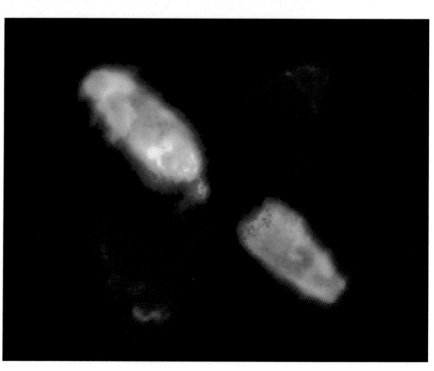

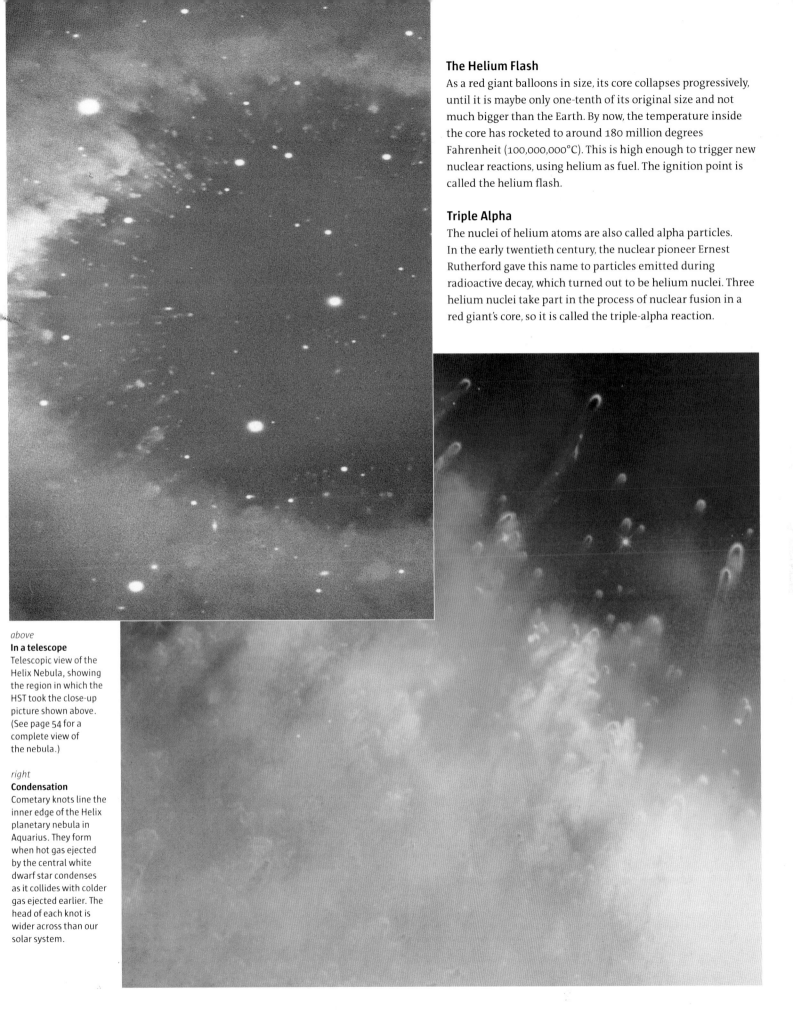

The Helium Flash

As a red giant balloons in size, its core collapses progressively, until it is maybe only one-tenth of its original size and not much bigger than the Earth. By now, the temperature inside the core has rocketed to around 180 million degrees Fahrenheit (100,000,000°C). This is high enough to trigger new nuclear reactions, using helium as fuel. The ignition point is called the helium flash.

Triple Alpha

The nuclei of helium atoms are also called alpha particles. In the early twentieth century, the nuclear pioneer Ernest Rutherford gave this name to particles emitted during radioactive decay, which turned out to be helium nuclei. Three helium nuclei take part in the process of nuclear fusion in a red giant's core, so it is called the triple-alpha reaction.

above
In a telescope
Telescopic view of the Helix Nebula, showing the region in which the HST took the close-up picture shown above. (See page 54 for a complete view of the nebula.)

right
Condensation
Cometary knots line the inner edge of the Helix planetary nebula in Aquarius. They form when hot gas ejected by the central white dwarf star condenses as it collides with colder gas ejected earlier. The head of each knot is wider across than our solar system.

In the process, first two helium nuclei fuse to form an unstable beryllium nucleus. If a third helium nucleus collides and fuses with that, it produces carbon. If a fourth collides and fuses, it produces oxygen.

Although all the hydrogen has been used up in the core, there is still hydrogen in the outer atmosphere. Temperatures are high enough immediately outside the core to cause the hydrogen to fuse. So the core takes on a kind of layered structure: in the center is accumulated carbon and oxygen; around that is an inner shell in which helium is fusing; and surrounding that is an outer shell of fusing hydrogen.

Stellar Winds

The heat produced in the core of a red giant rises to the surface on convection currents. These currents also carry a certain amount of core material, such as carbon, to the surface. The carbon condenses (turns into a solid) in the outermost layers, forming a soot-like dust.

As time goes by, gases in the outer atmosphere of the star gradually stream off into space as a stellar wind, just like the solar wind from the Sun. The star's pulsating—periodic expanding and contracting—sometimes exaggerates the process. The pulsations cause it to vary in brightness, becoming brighter as it contracts and heats up, dimmer as it expands and cools down.

The star Omicron Ceti in Cetus (the Whale) is a classic example of a pulsating red giant. At its brightest, it is second magnitude and is easily visible with the naked eye. At its dimmest, it has a magnitude of 10 and cannot be spotted even with binoculars. It varies from bright to dim, and back again, over a period of about 330 days. The star is usually known as Mira, which means "the wonderful." And the class of long-period variable stars it typifies are called Mira variables.

Dusty Eruptions

Usually, the stellar wind also carries with it the sooty dust grains in the outer atmosphere. Occasionally, however, the dust accumulates around the star and dims its light. Then a vigorous blast of stellar wind removes it, and the star resumes its former brightness. This happens with supergiants as well.

When we observe such a star, we see it as a variable with an irregular period, because the process of build-up and subsequent blowout can take weeks or even years. R Corona Borealis is a classic example. Most of the time it shines at a steady magnitude of six or so, on the edge of naked-eye visibility. But, unpredictably, it can suddenly dim to a magnitude of 12 or below. It may start to recover almost immediately or do so over the course of several months.

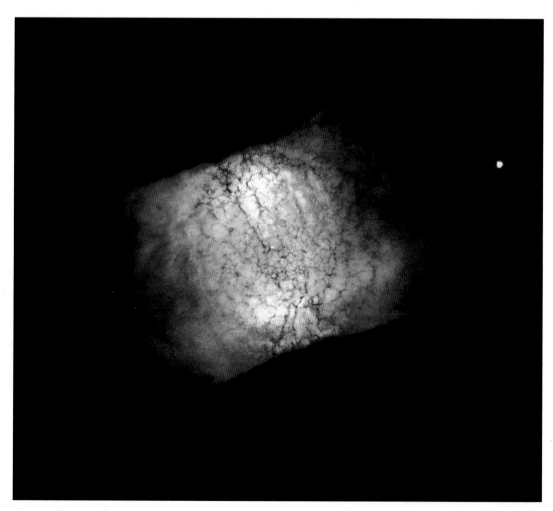

left
The Retina
Glowing in all colors of the rainbow, this planetary nebula (IC 4406) has a donut shape. It has been dubbed the Retina Nebula because the intricate tendrils of material we see have been compared with those in the eye's retina.

right
Cosmic ghost
This planetary nebula (NGC 6369) is known as the Little Ghost Nebula because it appears as a ghostly cloud around the faint star in its center. The faintest wisps of gas at the edge of the nebula were the first ejected by the central dying star. The inner prominent blue-green ring of material is nearly one light year in diameter.

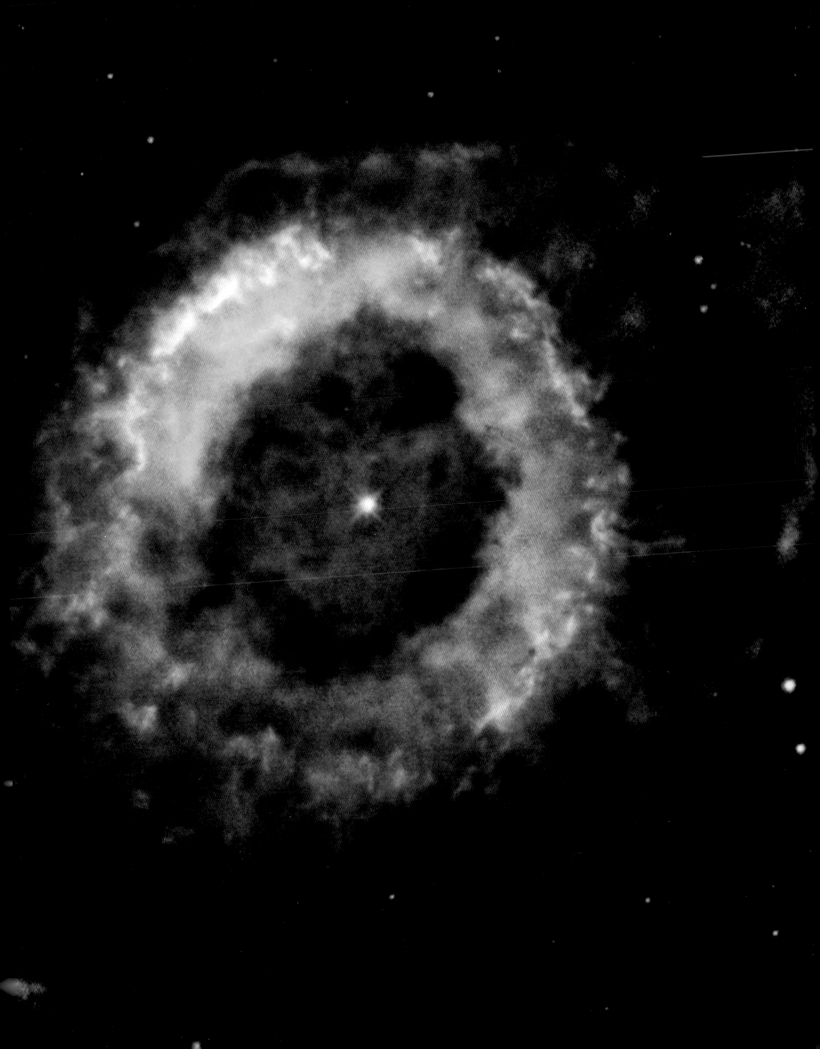

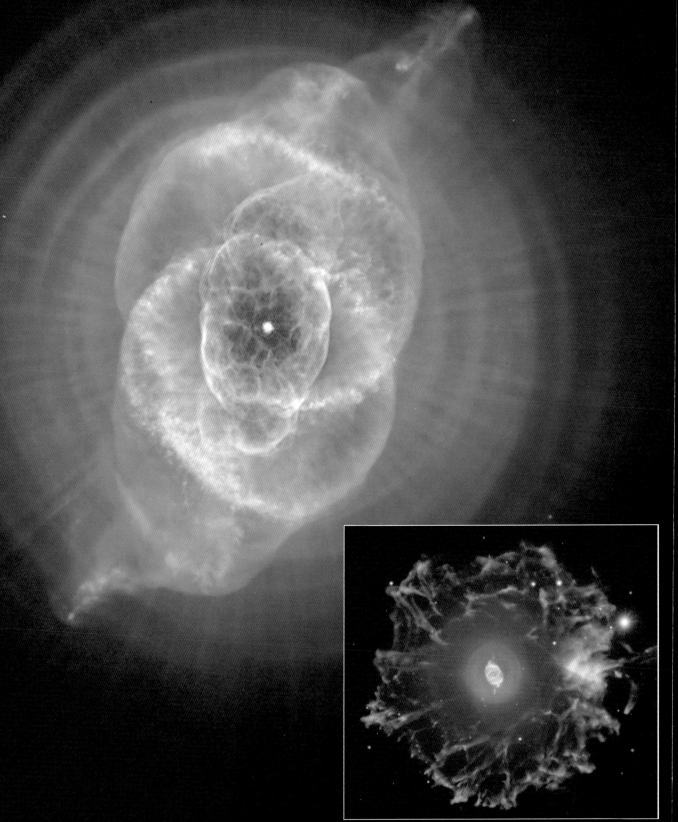

Last Gasps

Helium fusion in the core of a red giant lasts on average about two billion years. Then the helium runs out. Temperatures never get high enough to start fusing the carbon and oxygen that has been produced. In a star like the Sun, this marks the end of energy creation. The star is close to death.

Now come the final death throes. The super-hot core gives off a succession of shuddering death rattles that eject matter into space from the giant's outer layers. Traveling at velocities of up to 20 miles (30 km) per second, the successive waves of matter form into expanding shells. A torrent of X-rays and ultraviolet radiation streaming from the exposed core excite the ejected gas, and the shells light up with Technicolor brilliance.

Planetary Nebulas

Viewed through a telescope, these expanding shells are some of the most beautiful objects in the heavens. They are called planetary nebulas, although the term is misleading because they have nothing to do with planets. Eighteenth-century English astronomer William Herschel gave these objects their name because in the telescopes of the day they looked rather like the disks of planets.

Planetary nebulas take on a number of different forms. Some have the appearance of single or multiple spherical

bubbles, like M27, the Dumbbell Nebula in the constellation Vulpecula (the Fox). We see the bubble-like shells of other planetary nebulas as rings, such as the famous Ring Nebula in Lyra (the Lyre), profiled on page 54.

Of Butterflies and Cats

Most nebulas, however, are quite complex in shape. Many have overlapping rings of contrasting colors, caused by successive waves of ejected material punching through earlier ejecta. Others have an hourglass shape, with the ejected gas forming lobes on either side of the dying star. The Butterfly Nebula in Ophiuchus (the Serpent-Bearer) is a beautiful example.

The Cat's-Eye Nebula in the constellation Draco (the Dragon) has an amazingly intricate structure and thoroughly deserves its name. Eleven or more concentric shells of material make a layered onion-like structure around the dying star. One explanation for this layering is that the star ejected its mass in a series of pulses at 1500 year intervals.

With its extraordinary ability to distinguish detail, the HST has revolutionized our study of planetary nebulas. It reveals within these stellar ghosts a previously unsuspected complexity of structures and dynamic interactions.

left
The Cat's Eye
Well named, the Cat's Eye Nebula (NGC 6369 in Draco) is one of the most beautiful and complex planetary nebula the HST has studied. The eye is surrounded by concentric rings, which are spherical bubbles of material ejected at 1500 yearly intervals.

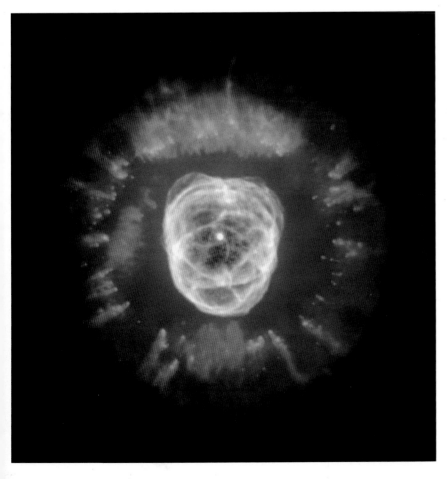

left
The Eskimo
The HST's exquisite rendering of the Eskimo planetary nebula (NGC 2392 in Gemini), so called because when viewed in ground-based telescopes it resembles a face surrounded by a fur parka.

above
The Egg Nebula
Dust layers that resemble onion skins surround a dying star. Twin beams of light from the hidden star illuminate the dust. The HST image of the Egg Nebula has been artificially colored so astronomers can see how the light reflects off the smoke-sized dust particles.

following pages
Southern Ring
Planetary nebula NGC 3132, in Vela, is well named the Southern Ring. It rivals the northern Ring Nebula in Lyra in beauty. The white dwarf ejecting material is the fainter of the two stars in the center.

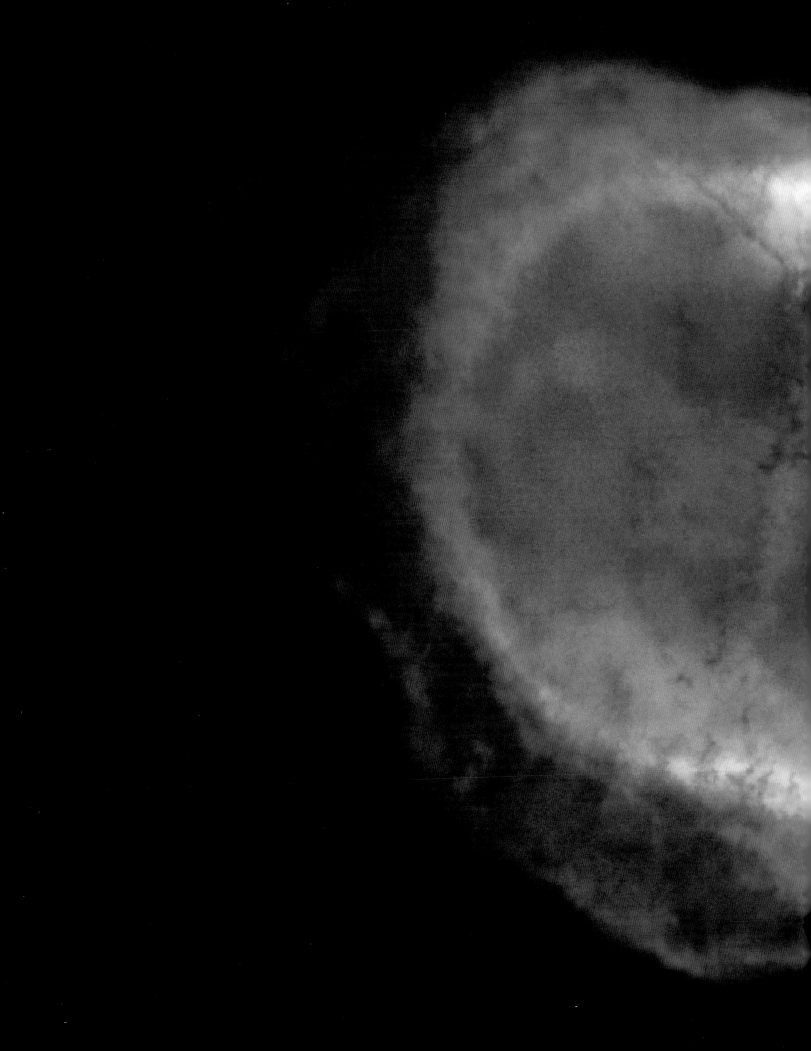

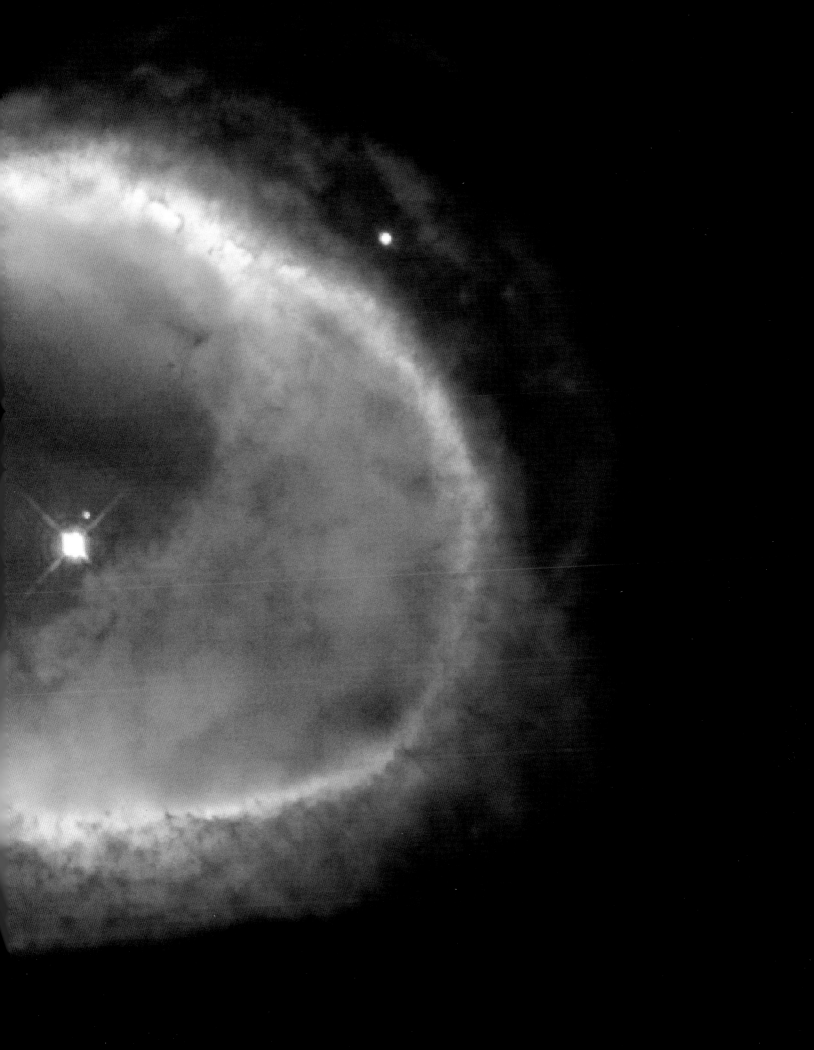

Heavyweight Dwarfs

The outflow of matter in stellar winds and pulsations can deplete a red giant of as much as 80 percent of its mass. In tens of thousands of years, the nebula will disperse into the general interstellar medium and disappear. All that will then remain is the exposed core.

Once the core runs out of helium, no energy-producing processes can occur there. Temperatures inside are not high enough to trigger a new round of nuclear-fusion reactions with the carbon and oxygen that make up the core.

With no radiation, the core can no longer resist the inward pull of gravity, so it progressively collapses. Eventually it shrinks to about the same size as the Earth and becomes a class of star we call a white dwarf.

The gravitational energy released as the core shrinks manifests itself as heat. The temperature of the surface of the white dwarf soars to tens of thousands of degrees and beyond. The HST has spotted a white dwarf in the planetary nebula NGC 2440 with a surface temperature of 360,000 degrees Fahrenheit (200,000°C). It is one of the hottest stars we know, nearly 40 times hotter than the Sun.

Degenerate Matter

With the mass of the Sun squeezed into a volume the size of the Earth, a white dwarf is extremely dense. Just a teaspoonful of its matter would weigh tons.

Furthermore, the matter that makes up a white dwarf is not ordinary matter. Ordinary matter consists of atoms made up of nuclei, with swarms of electrons circling round them at a distance. In a white dwarf, the matter is crushed. The electrons are practically pressed up against the nuclei. The repulsion between the electrons, which are all negatively charged, prevents the body from collapsing further. This kind of matter is called electron-degenerate matter.

Chandrasekhar's Limit

Typically, the core of a white dwarf has about the same mass as the Sun. It can never have more than 1.4 times the mass of the Sun—a surprising discovery made in 1930 by Indian astronomer Subrahmanyan Chandrasekhar. If the core of a dying star has more than 1.4 solar masses (called the Chandrasekhar limit), gravity overcomes the electron repulsion and causes further collapse (see page 62).

The Final Curtain

What eventually happens to a white dwarf? Most of them just quietly fade away. As they radiate away their remaining energy, they cool and dim, and their light reddens. After many billions of years they run out of energy completely. They become black dwarfs and disappear from the visible universe. Scientists are uncertain whether any black dwarfs exist. It may be that the universe is not yet old enough.

The Pup

The first white dwarf that astronomers discovered was the so-called Companion of Sirius, the Dog Star. It is nicknamed the Pup. (Astronomers call it Sirius B, and the Dog Star Sirius A.) In 1834, the German astronomer Friedrich Bessel noticed that Sirius, which is one of the nearest stars, pursues a rather erratic course through the heavens. It weaves back and forth against the background of more distant stars.

Bessel concluded that this erratic behavior must be caused by an orbiting companion star, but he still hadn't found it when he died in 1844. US astronomer Alvan Clark was the first to spot the companion star 18 years later.

With a magnitude of about eight, the Companion of Sirius is theoretically bright enough to be visible in binoculars, but it gets lost in the glare of Sirius itself, brightest of all the stars.

right main image
The Bug Nebula
This planetary nebula is one of the brightest and most extreme known. The dying star that expelled the surrounding wings of material is hidden from view. It is within a dusty torus of gas (upper right). The star with a temperature of at least 450,000°F (250,000°C) is one of the hottest known.

right inset
Blue skies
This glorious 2003 Hubble Heritage release reveals glowing knots of gas inside the Dumbbell Nebula. They seem to be a common feature in planetary nebulas. Here, the blue color indicates concentrations of oxygen gas.

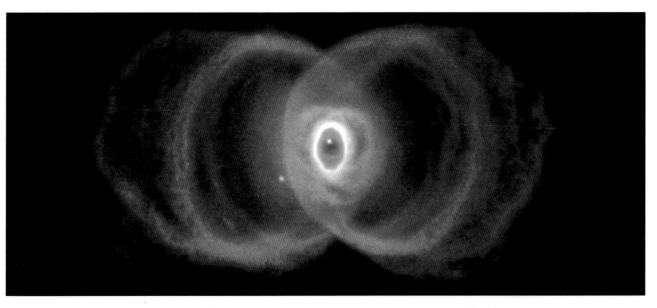

left
In the Hourglass
The white-dwarf star MyCn 18 lies at the heart of the Hourglass Nebula, some 8,000 light-years away. Note that it lies off-center—this suggests that it may have a faint companion.

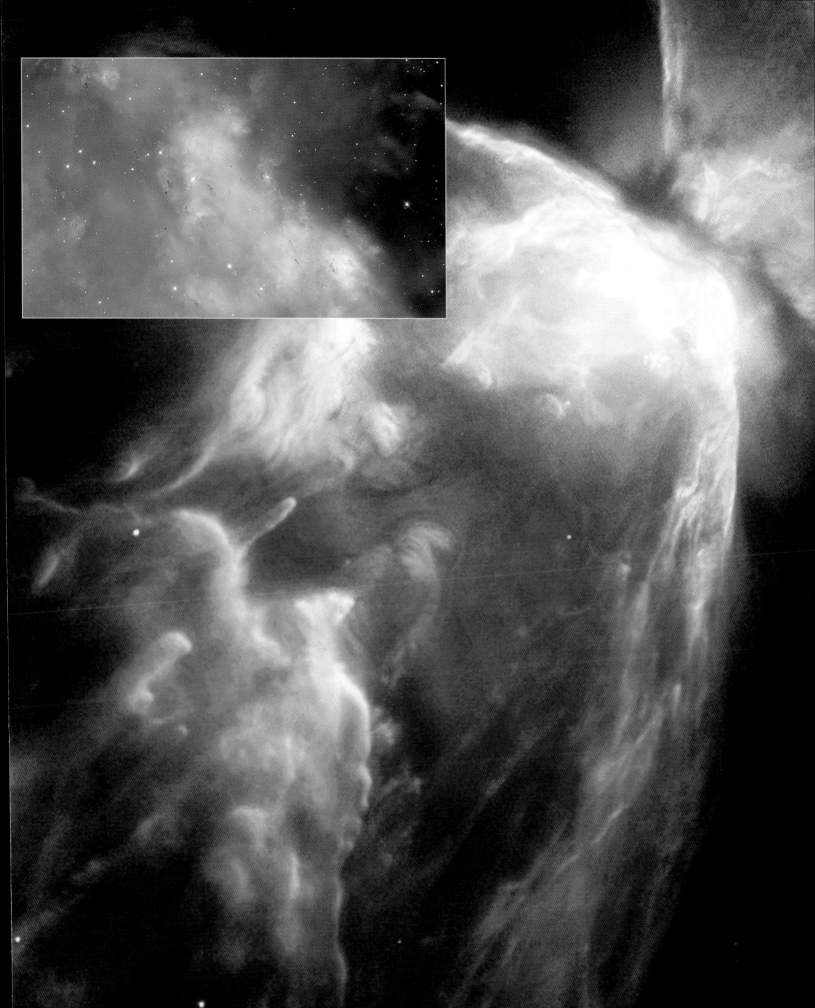

The Ring Nebula

Lyra (the Lyre) is one of the smallest constellations in the heavens. Greek astronomers named it for the lyre that Hermes, the messenger of the gods, made out of a tortoise shell strung with cow gut.

This tiny constellation has two outstanding features. One is its lead star Vega, the fifth-brightest star in the sky and one of the three bright stars that blaze overhead during the northern summer. The others are Deneb in the adjacent constellation Cygnus (the Swan) and Altair in Aquila (the Eagle). Together they form the famous Summer Triangle.

The other outstanding feature in Lyra is M57, the Ring Nebula. It is a planetary nebula — in fact, the definitive planetary nebula before the HST revealed the staggering beauty of so many others.

The Ring Nebula is easily located because it is sandwiched between Beta and Gamma, the second- and third-brightest stars in the constellation. It is not visible with binoculars, but even small telescopes reveal its "planetary" shape and smoke-ring appearance. Larger instruments are needed to show its central blue-white star, which is a white dwarf. It is one of the hotter types, with a surface temperature of around 180,000 degrees Fahrenheit (100,000°C).

Like other planetary nebulas, the Ring Nebula is growing as the shell of material ejected by the central star pushes further into space. The outer edge of the ring is well defined, which makes it possible to calculate its rate of expansion as close to 12 miles (20 km) per second. Working backwards, this gives an estimated date for the origin of the nebula of around 5,500 years ago.

Several other planetary nebulas have the appearance of cosmic smoke rings, although they are not as perfect as the Ring Nebula itself. One is the Helix Nebula in Aquarius (the Water-Bearer). It has two overlapping rings, which make it look like a spiral or helix.

right
Captivating Ring
Seen in exquisite detail by the HST, the Ring Nebula appears to have a cylindrical, or hourglass shape. We see it as a ring because we happen to be looking at it end-on.

below left
Stellar smoke ring
With a magnitude of around 9, the Ring Nebula can be clearly viewed through small telescopes. Roughly 1 light-year across, it lies about 2,000 light-years away.

below right
The Helix
This view of the Helix Nebula in (NGC 7293) is a composite of HST and ground-based images. It appears to be a simple ring shape but looks are deceiving. If viewed from the side, two gaseous disks nearly perpendicular to each other would be visible.

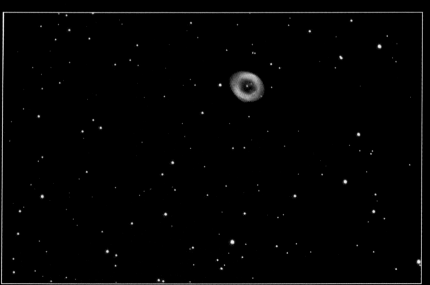

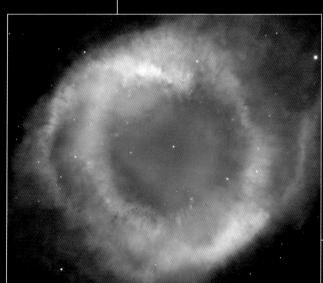

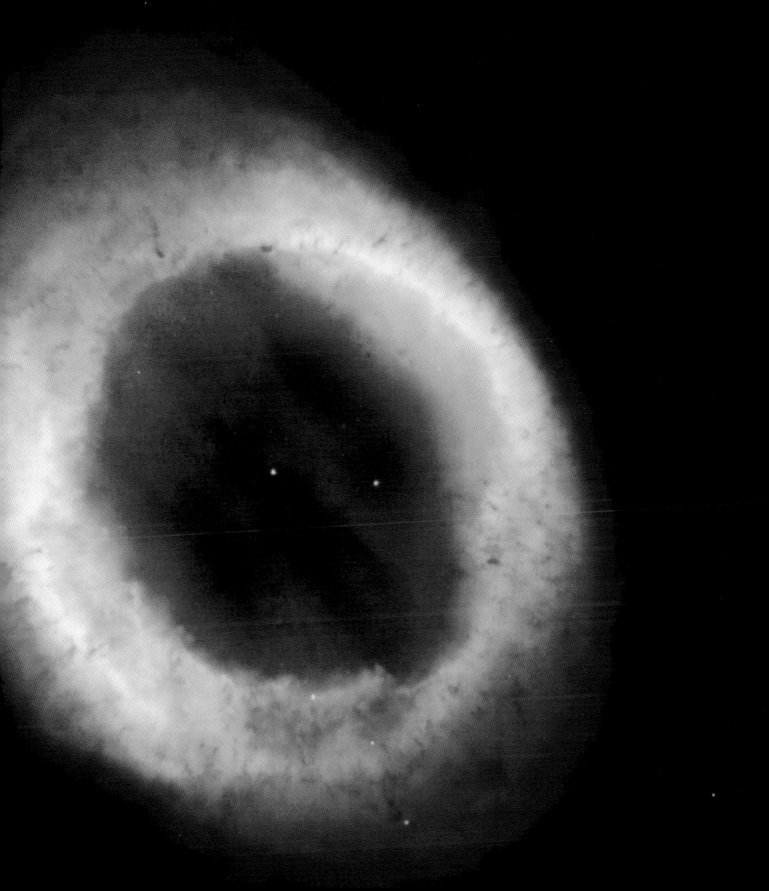

New Stars

Sometimes a white dwarf forms part of a binary, or two-star system, along with a larger star. It then starts to gain gas—mostly hydrogen—from its companion. As more and more hydrogen rains down, a thick, dense layer builds up, and temperatures and pressures on the white dwarf begin to soar.

Eventually, conditions are extreme enough to trigger hydrogen fusion. A massive star-wide thermonuclear explosion blasts material into space and makes the star flare up to perhaps 100,000 times its original brightness in a few days. Over the following weeks it returns to its original state, until the same process begins all over again.

Such sudden flare-ups are called novas, which means "new." Early astronomers gave this name to what seemed to be new stars appearing in the heavens, when in fact they were existing stars that had been faint but suddenly became visible.

Supernova

Some "new stars" flare up to an even greater brilliance. They can increase millions of times in brightness and become as bright as a whole galaxy. We call these flare-up stars supernovas. On average, one supernova occurs in a galaxy about every 200 to 300 years.

The first recorded supernova seems to be the one that Chinese astronomers noted in 1054 in Taurus. We see its ghostly remains today as the Crab Nebula. The Danish astronomer Tycho Brahe spotted another supernova in our Galaxy in 1572, when it became as bright as Venus and was visible in daylight. It became known as Tycho's Star. In 1604, Tycho's protégé, the German astronomer Johannes Kepler, saw another supernova, which has been dubbed Kepler's Star.

That was the last time a supernova was seen in our Galaxy, although many have been spotted in others. Most notable was one in our galactic neighbor the Large Magellanic Cloud in 1987. Not yet operational at that time, the HST couldn't witness the event, but it has been used to study carefully the aftermath ever since.

We call the visible remains of a supernova explosion (such as the Crab Nebula) a supernova remnant (SNR). We find many other examples in the heavens, which often rival planetary nebulas in their delicate beauty.

Types I and II

There are two kinds of supernovas. A Type I supernova occurs in the same binary big-star/white-dwarf system that creates a nova. It is created when the white dwarf gains so much matter from the larger star that it can no longer support itself. It collapses and destroys itself in a supernova explosion.

A Type II supernova occurs when a star with much greater mass than the Sun dies. Type II supernovas are more common. They release much more energy than Type Is, but not in visible light. Type I supernovas appear brighter.

left
Stellar explosion
These wisps of glowing gas were expelled by a star 3,000 years ago as it went supernova. The titanic explosion was in the nearby galaxy, the Large Magellanic Cloud. The complex structure of this supernova remnant, N132D is revealed by combining HST and images taken by the Chandra x-ray space telescope.

right
Making waves
The supernova explosion that occurs when a massive star blasts itself apart sends powerful shock waves rippling through the surrounding space. These waves compress interstellar gas and make it glow, creating the luminous filaments we see in this image.

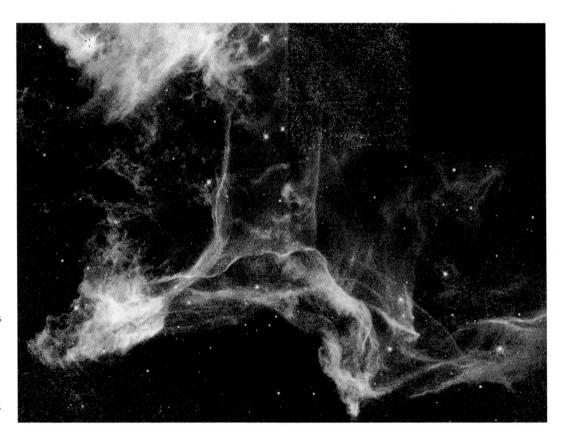

Eta Carinae

The HST has been used to examine closely one supermassive dying star: Eta Carinae, in the far southern constellation of Carina (the Keel).

Eta Carinae is one of the biggest, hottest and intrinsically brightest stars we know. Compared with the Sun, it is 100 times more massive, 150 times wider, and four million times more brilliant. It is also more than five times hotter than the Sun, with a surface temperature nudging 54,000 degrees Fahrenheit (30,000°C).

Eta Carinae has a visual magnitude of about seven, which puts it just beyond naked-eye visibility. But it is easily spotted with binoculars and a small telescope, and looks noticeably colored.

Astronomers know that a star with such staggering statistics must be inherently unstable, and so it has proved. Eta Carinae periodically flares up and then dims, fluctuating in brightness over irregular periods of time.

English Astronomer Royal Edmond Halley (of comet fame) watched the star brighten to the fourth magnitude in 1677. An even more spectacular flare-up began in 1835 and peaked in 1843, when Eta Carinae became the brightest star in the heavens after Sirius. Since then it has gradually dimmed.

Eta Carinae, then, is a variable star. It is called an eruptive variable because it varies in brightness when it undergoes massive eruptions from its atmosphere that blast matter into space. It flares up because the ejected matter is very hot and radiates abundant light.

Vast amounts of matter were ejected in Eta Carinae's massive eruption in the 1840s. Today we see the ejecta as an expanding red shell of nitrogen, oxygen and other gases, traveling at speeds up to two million miles per hour (3,000,000 km/h).

The violent eruption Eta Carinae suffered is reminiscent of a supernova explosion, but it was only a hint of what is to come. When this doomed star really goes supernova and blows itself to bits — which could happen any time — it will be a truly amazing sight.

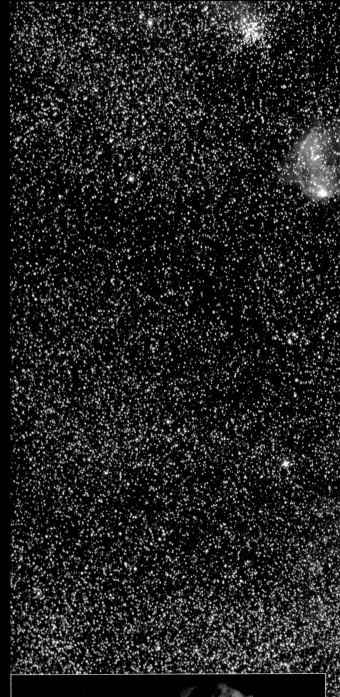

right (main image)
Wide angle
A glorious stellar vista featuring the Eta Carinae Nebula is captured in this wide-angle photograph taken by a ground-based Schmidt telescope.

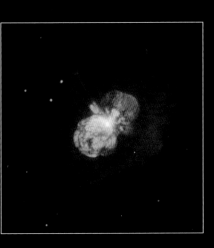

left
Flare-up
The HST has peered into the nebula to view the vast clouds of gas and dust ejected by the star Eta Carinae some 150 years ago. That was when it flared up to become one of the brightest stars in the heavens.

right inset
Doomed superstar
This later image shows spectacular detail of the massive clouds of ejected matter. It even glimpses the blue light coming from the intensely hot hidden star.

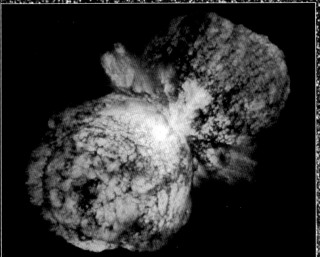

The Death of Massive Stars

Massive stars destroy themselves in a Type II supernova explosion. As was noted earlier, a star with about the same mass as the Sun rides the main sequence for billions of years, swells up into a red giant, puffs off its outer layers and then shrinks to become a tiny white dwarf.

But stars much more massive than the Sun live life in the fast lane. They consume their hydrogen fuel voraciously and spend only a few tens of millions of years on the main sequence. The biggest stars of all may spend less than a million years there.

As with a star like the Sun, a supermassive star stays on the main sequence while hydrogen fuses into helium in its core. When the hydrogen runs out, it starts swelling up, becoming cooler and redder. Because it is so massive, it swells up beyond the size of a red giant to become a supergiant, with hundreds of times the diameter of the Sun.

Inside a Supergiant

Like in a red giant, the core of a supergiant starts collapsing when it runs out of hydrogen. The collapse raises the temperature and initiates the fusion reactions that turn the helium into carbon and oxygen.

Once all the helium has been used up, the core begins to collapse again. Because the core in a supergiant is so massive, its collapse releases prodigious amounts of energy, pushing temperatures up to hundreds of millions of degrees. This is high enough to make the carbon and oxygen fuse to produce heavier elements, such as neon, magnesium, silicon and sulfur. These elements fuse in turn. The final product is iron, which is stable and will not fuse.

The rapidity with which these elements fuse is staggering. The carbon "burn" may take only about a thousand years, the oxygen burn just a year and the final silicon burn to form iron only a few days.

The Mother of all Explosions

Once the entire core of a supergiant has been converted into iron, its enormous mass causes it to collapse. The collapse is so rapid that the rest of the star crashes down on itself as well. The release of energy is catastrophic and blasts the star to smithereens. It becomes a supernova.

The temperatures and pressures produced in this cataclysmic event convert iron into a succession of heavier elements. In fact, all the elements heavier than iron that exist in nature were forged in supernovas. These mothers of all explosions blast newly made elements into interstellar space, where they make their way into the vast nebulas that spawn new stars.

How neat this is: the spectacular death of massive stars seeds the universe with the material to make new ones — celestial recycling.

Supernova SN 1987a

On February 23, 1987, astronomers spotted a bright "new star" in southern skies. It was a brilliant supernova, designated 1987a. Over the next weeks it nearly reached the second magnitude of brightness and was easily visible to the naked eye. It was the brightest supernova seen since Kepler's Star of 1604. But that star was not in our Galaxy; SN 1987a was in a neighboring galaxy, the Large Magellanic Cloud, some 160,000 light-years away.

The star that went supernova was identified as Sanduleak −69°202. It was a blue supergiant estimated to have had a mass 20 times that of the Sun. Shortly after the HST went into orbit in 1990, it spied a ring of matter that had been blasted into space by SN 1987a. It has been charting the incredible convolutions of this expanding ring ever since.

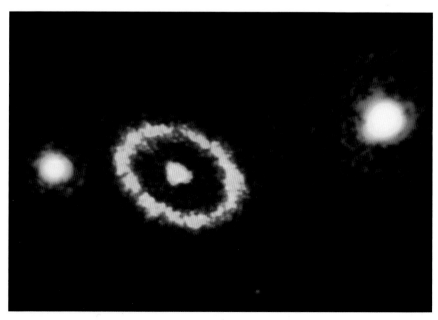

End of the Line

Exactly what happens to the collapsing core of a massive star in the aftermath of a supernova explosion depends on its mass. It will either become a neutron star or that most awesome of celestial objects—a black hole.

If the core has up to about three times the mass of the Sun, it becomes a neutron star. The force of collapse creates a new kind of matter. It forces the electrons in atoms into the nucleus, where they merge with the protons and turn into neutrons. The core becomes a mass of neutrons, tightly compressed together—a form of matter called baryon degenerate matter. (Neutrons belong to a class of subatomic particles called baryons.)

Such a neutron star is tiny, averaging only about 12 miles (20 km) across. Yet it contains three or more solar masses. So it has the most incredibly high density—millions of times greater even than a white dwarf.

Pulsating Signals

After a supernova explosion, the neutron star is left spinning furiously. Its powerful magnetic field spins as well, and funnels the star's energy into powerful beams of radiation that emanate from the two magnetic poles.

As the star rotates, these beams sweep around in space, rather like those from a lighthouse. If the beams are in our line of sight, we detect pulses of energy every time they sweep past. We call these pulsating bodies pulsars.

Little Green Men

Working at Britain's Cambridge Radio Astronomy Observatory, astronomer Jocelyn Bell-Burnell discovered the first pulsar in 1967 by its radio waves. Never before had such a regularly pulsating radio source been detected. Initially baffled over what it could be, the Cambridge astronomers dubbed it LGM (for Little Green Men), likening it to the kind of signals an alien civilization might beam into space.

Since then, more than 1,500 pulsars have been discovered. Most give off energy as radio waves, but some give off energy in other forms, including visible light, X-rays, and gamma rays.

The fastest pulsar, PSR J1748-2446ad, was discovered in January 2006. It is in a globular cluster of stars called Terzan 5, in the constellation Sagittarius and is spinning 716 times per second. It whirls round faster than the blades of a kitchen blender. Try to imagine something the size of New York spinning around that fast!

Black Holes

After a supernova, if the collapsing core has a mass greater than about three solar masses, it becomes a black hole. When such a massive core collapses, the powerful gravitational forces compress even neutrons, causing the body to go beyond the neutron-star stage. As the core gets ever smaller, the surface gravity gets ever larger, and the velocity needed for anything to escape from it escalates.

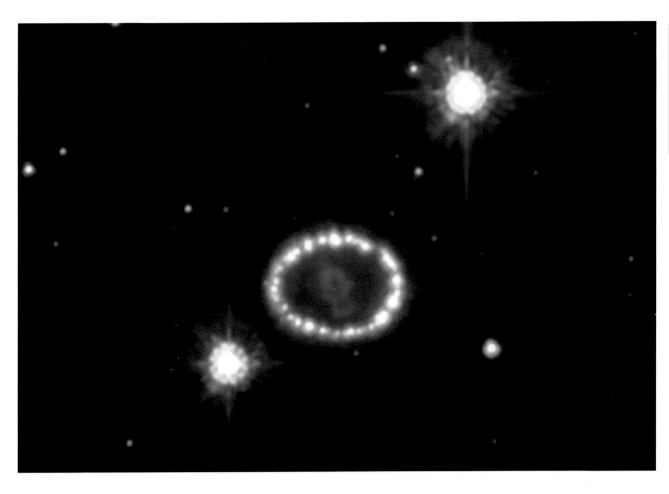

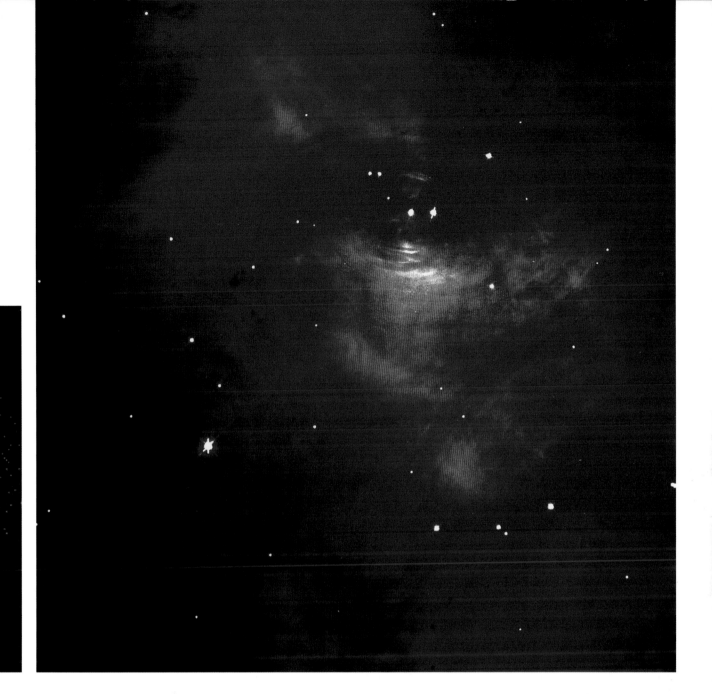

Eventually the surface gravity of the collapsed core becomes so high that not even light can escape from it. At this point it becomes a black hole and disappears from the observable universe. The circle that marks the edge of a black hole — the boundary between it and the visible universe — is called the event horizon. Its radius, called the Schwarzschild radius, depends on the mass of the collapsing body. Usually it measures just a few miles.

What lies within the event horizon is, of course, impossible to fathom. The theory is that the collapsing core continues to shrink until it becomes an infinitely small point of infinite density known as a singularity. The reality could be different — we will probably never know.

X Marks the Spot

Looking for black holes in the blackness of space is obviously impossible, but there are ways to detect them. When a black hole exists in a binary, two-star system, it attracts gas from the other star with its powerful gravity. As the two bodies orbit each other, the gas first swirls round the black hole, forming what is called an accretion disk.

The material in the disk swirls round very fast, and becomes searing hot because of friction between the particles. It gives off energy as X-rays before spiraling into the black hole. And we are able to detect these X-rays.

Cygnus X-1, a powerful X-ray source in the constellation Cygnus, is believed to come from such a binary black-hole system. It was the first object that astronomers suspected was a black hole, and they estimate that it contains the mass of nearly 20 Suns.

There are some black holes that have a mass that is millions of times the mass of the Sun. These supermassive black holes are found in the centers of galaxies, where they were produced by the collapse of enormous gas clouds, rather than by dying stars. They are the extremely powerful "engines" that are responsible for the extraordinary energy output of quasars and other active galaxies (see page 78).

above
In the Crab
The HST has homed in on the heart of the Crab Nebula to reveal the pulsar remnant of the 1054 supernova explosion. The pulsar is the left star of the bright pair in the center. It measures only about 6 miles (10 km) across and spins around 30 times a second.

3 | Gregarious Galaxies

The HST reveals the beauty of these great star islands in space

above
Starburst galaxy
Young stars are being born ten times faster in the central region of the starburst galaxy M82 than inside the Milky Way. Superwinds from hot, bright stars help form towering plumes of hot gas, seen here in red, above and below the disc of the galaxy.

inset left
Barred spiral
The HST's sharp eye has revealed myriad fine details in barred spiral galaxy NGC 1300. Blue and red supergiants are seen in the spiral arms, and dust lanes trace out fine structures in the disc and central bar. Numerous more distant galaxies are seen in the background.

inset right
Edge-on
The galaxy NGC 5866 is tilted nearly edge on to our line of sight. Face on it would show a flat disc with spiral structure. Edge on we see a bright nucleus split by a dust lane. A blue disc of stars runs parallel to the dust lane, and a transparent outer halo dotted with globular clusters surrounds it all.

The Milky Way

In whichever direction you look in the night sky, you see twinkling stars, scattered in the velvety blackness of space. So is this what the universe is like—stars scattered haphazardly in space? Indeed it is not. If you could take a ride on an interstellar spaceship, head for the stars and then keep going some, you would eventually leave the stars behind. Looking back, you would see that all the stars in the sky are grouped together, forming a great star island in a vast ocean of space. And looking in other directions, you would see other star islands, with empty space in between. We call these separate star systems galaxies. Edwin Hubble was the first to prove the existence of external galaxies beyond our own and study them in detail.

Our own galaxy is called the Milky Way, but we often also refer to it simply as "the Galaxy." Like most other galaxies, it is a collection of billions of stars, bound together by gravity. It has a spiral shape and rotates slowly. From a distance it would look like a spinning wheel-shaped firework. Many galaxies are similar to it, but others are different. The most extraordinary ones pump out unbelievable amounts of energy. They include quasars—some of the most amazing and remote objects in the universe.

In the study of galaxies, the HST reigns supreme. With its incomparable clarity of vision and sensitivity, it provides panoramic vistas of deep space showing galaxies by the hundreds. It can also zero in on individual star systems to tease out the most minute details.

Look at the heavens on a clear, moonless night, and you'll see a faint, misty band of light arching across the sky. In the Northern Hemisphere, it runs through the constellations Cassiopeia, Perseus, Cygnus and Aquila, and in the Southern Hemisphere, Crux, Centaurus, Scorpius and Sagittarius.

What is this tenuous misty band? The ancient Greeks said it was a stream of milk spurting from the breast of the goddess Hera, the long-suffering wife of Zeus, the philandering king of the gods. They called it Kiklos Galaxias, meaning the milky circle. We know it today as the Milky Way. It was the Italian Galileo who first discovered the nature of the Milky Way, when he turned his telescope on the heavens in the winter of 1609-10. He saw that it was made up of countless numbers of faint stars, seemingly packed tightly together. If you look at the Milky Way through powerful binoculars, you'll see that he was right.

Later the following century, astronomers such as William Herschel began to realize what the Milky Way really is. It is a view from inside a layer of stars that form our star system— our Galaxy. By counting stars in regions of the sky on either side of the Milky Way, Herschel concluded that the Galaxy was lens-shaped: thickest at the center and thinnest at the edge. He was not far wrong.

Our Galactic Home

Our Galaxy, also called appropriately the Milky Way, is shaped somewhat like a disk with a huge bulge in the middle. Long, curved arms spiral out from the bulge, or nucleus. Rather inelegantly, the shape of the Galaxy has been likened to that of two fried eggs, stuck back to back!

The size of the Galaxy is astonishing. It measures 100,000 light-years from edge to edge. The nucleus is about 6,000 light-years thick, while the disk averages only about a third of this.

The number of stars it contains is uncertain; it is thought to have at least 500 billion stars. The Sun is located on one of the spiral arms, about 25,000 light-years from the center. The center lies in the direction of the constellation Sagittarius. And this is where the Milky Way appears brightest in the sky.

right
Face-on
Fifty-one individual HST exposures have been combined to produce this stunning image of the gigantic Pinwheel Galaxy (M33 in Triangulum). It is twice the diameter of the Milky Way Galaxy and contains at least one trillion stars but gives an idea of what we'd see if we could view our galaxy from afar.

right
Night sky view
Small telescopes show the Milky Way well, revealing a dazzling panorama of close-packed stars, bright clusters and colorful nebulas. Many dark dust lanes are evident, particularly in the far southern constellations.

In the Center

The central bulge of the Galaxy contains mostly old red and yellow stars. There is comparatively little interstellar matter, and star formation is limited.

Until the last decades of the 20th century, what lay in the interior of the bulge was a mystery, because thick clouds of gas and dust obscure our view. Radio and infrared studies of the region have now revealed a host of fascinating features.

At the exact center is a powerful radio source known as Sagittarius A. It appears to mark the site of a black hole with the mass of 2.5 million Suns. Astronomers believe that supermassive black holes like this might lurk at the center of most galaxies. Magnetism is also a dominant force in the galactic center, creating strange structures such as the Arc.

The Spiral Arms

Using radio telescopes to penetrate the obscuring clouds in the Milky Way, astronomers have created a detailed map of the Galaxy. These radio surveys have revealed the curved, spiral "arms" that make up the disk. Young, bright stars and bright and dark nebulas make up the shape of the arms.

There are two main spiral arms, Sagittarius and Perseus, as well as pieces of others. Our own Sun is on the Orion Arm, which forms a kind of bridge between the Sagittarius and Perseus Arms. Riding the arm in our neighborhood are stars of some of the most familiar constellations, such as Orion, Taurus and Cygnus. Clusters such as the Hyades and Pleiades lie relatively close, as do some of our most familiar nebulas, such as the Orion, Helix, Dumbbell and North America.

Farther away from the galactic center, the Perseus Arm is quite fragmented. It contains several well-known supernova remnants, including the Crab Nebula and Cassiopeia A. The Rosette Nebula in Monoceros and the double star cluster in Perseus are among other delights.

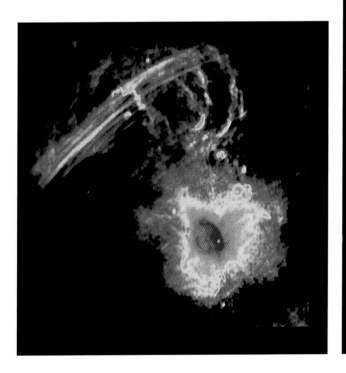

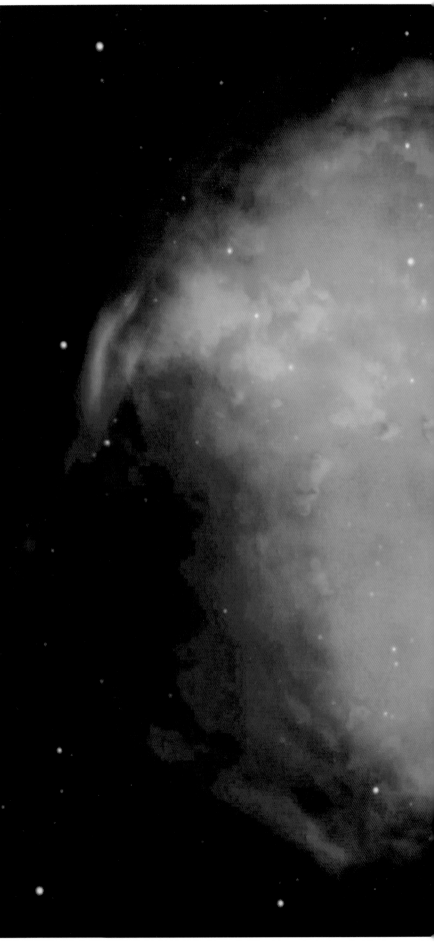

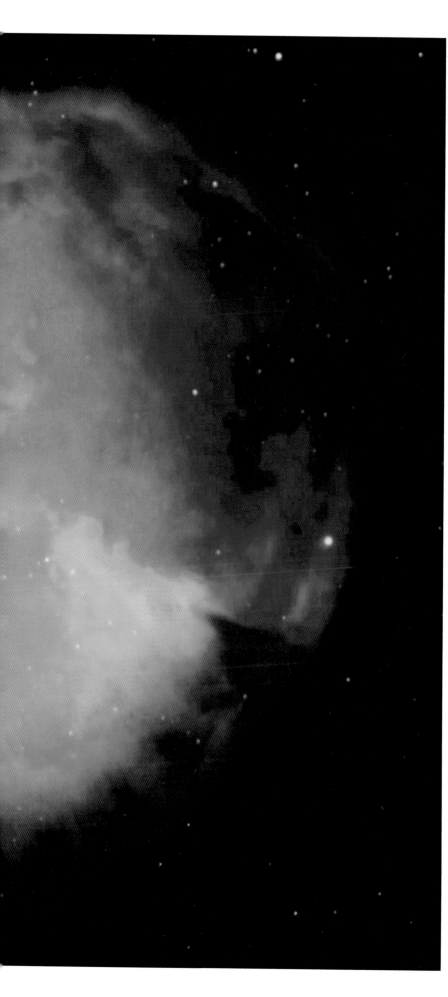

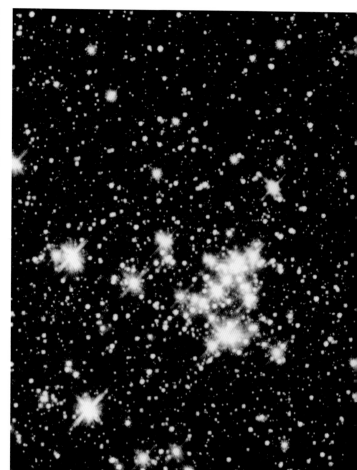

far left
On the radio
This radio image of the center of the Galaxy shows the curved feature called the Arc, which seems to be part of a ring of highly magnetized gas (the radio lobe). The white point is the radio source Sagittarius A.

left
On the Orion Arm
The Dumbbell planetary nebula is one of the many familiar night-sky objects in our local celestial neighborhood. It lies about 1,000 light-years away. Its dumbbell shape is more evident in less powerful telescopes than here.

above
Quintuplets
Near the center of the Galaxy, the HST has spotted a group of giant stars, called the Quintuplet Cluster. Ten times larger than typical clusters in the spiral arms, it is the home of the brightest star we know in the Galaxy, the Pistol Star (see page 177).

But it is on the Sagittarius Arm, closer in to the galactic center than the Orion Arm, that we find some of the most spectacular night-sky objects. Great billowing clouds of interstellar matter mark regions of intensive star birth in the part of the arms closest to us. They include the Eagle, Omega, Trifid and Lagoon Nebulas. The unstable giant star Eta Carinae threatens to go supernova, while the sparkling stars in Crux's Jewel Box cluster could not be more appropriately named.

On the Outside

Most of the matter in the Galaxy resides in the stars in the bulge and spiral arms, and in the interstellar medium—the gas and dust between the stars. But a few hundred tightly-knit globular star clusters circle outside the bulge and disk, and there are also large amounts of invisible dark matter within a great spherical halo that envelops the entire Galaxy.

Classifying the Galaxies

Our home Galaxy—a rotating spiral of hundreds of billions of stars that measures thousands of light-years across—is one of billions of such galaxies in the universe. They are known as spirals. Other galaxies consist only of a spherical or elliptical bulge of stars and lack the curved arms of spirals. They are known as ellipticals.

Spirals and ellipticals are termed regular galaxies because of their distinctive shape. But some galaxies have no particular shape and are termed irregulars. Edwin Hubble devised the method of classifying regular galaxies according to their shape in his so-called tuning fork diagram.

Elliptical Galaxies

More than half of all galaxies are ellipticals. They consist mainly of old stars and contain little free gas or dust, and so little or no star formation takes place within them. Elliptical galaxies vary widely in size and mass. The largest and some of the smallest galaxies are ellipticals. The giant elliptical galaxies at the heart of galaxy clusters measure many hundreds of thousands of light-years across.

At the other end of the scale are dwarf elliptical galaxies just a few hundred light-years across. Many of our galactic neighbors are dwarf ellipticals, some containing only a few hundred thousand stars, rather than billions.

In the Hubble classification, ellipticals are denoted E, followed by a number from one to seven describing how round or oval they are. E0 is a near-spherical galaxy and E7 the most flattened oval.

Spiral Galaxies

Spirals have essentially the same structure and composition as the Milky Way, with a flattened bulge at the center, and stars and dust carried on arms spiraling out. The bulge contains mainly older stars, with little gas and dust. It is orbited by a few hundred globular clusters. The arms, by contrast, are populated mainly by young stars, and star formation is rampant among the turbulent gas clouds of the interstellar medium.

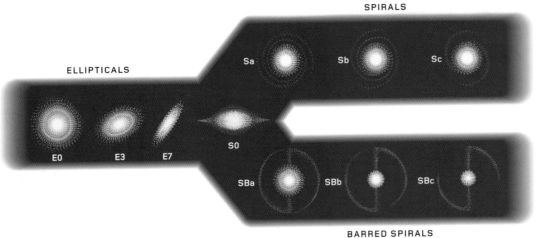

SPIRALS

ELLIPTICALS

Sa   Sb   Sc

E0   E3   E7   S0

SBa   SBb   SBc

BARRED SPIRALS

left

Hubble's tuning fork
Edwin Hubble's system of classifying regular galaxies into ellipticals, spirals and barred spirals.

above

Dusty spiral
A beautiful spiral galaxy (NGC 4414) in Coma Berenices, which has copious amounts of dust in its spiral arms. As is usual in spirals, the central bulge has older, yellower stars, while the arms have many younger, bluer ones.

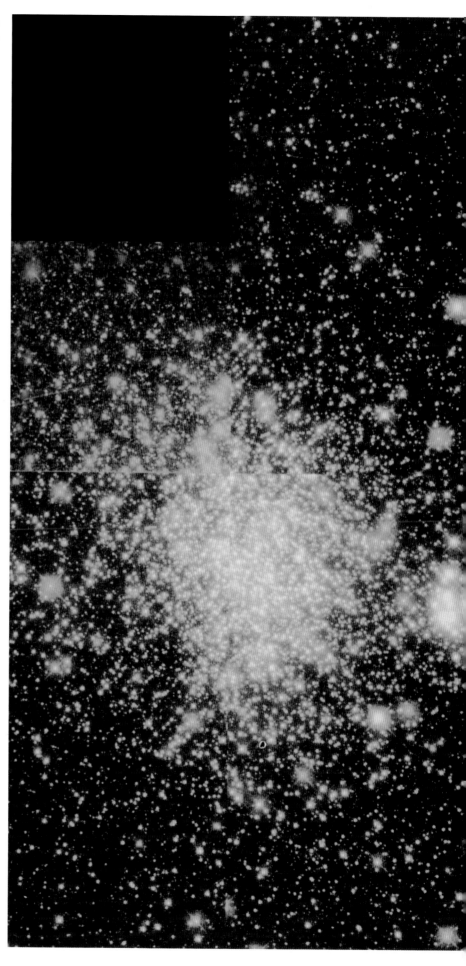

above
Bar of stars
The nearby galaxy NGC 1672 is a prototypical barred spiral galaxy. Its spiral arms are attached to the two ends of a straight central bar of stars. At either end of the bar are regions of concentrated star formation.

right
Seeing double
The HST reveals that the cluster NGC 1850, in the Large Magellanic Cloud, is a double. The dominant cluster is made up of yellowish stars around 40 million years old. The scattered white-hot stars in the other cluster are ten times younger.

Spirals do not vary as much in size as ellipticals. The Milky Way seems to be of about average size. In our galactic neighborhood, the galaxy in Triangulum is less than half its size, while the Andromeda Galaxy (see page 102) is half as big again.

Perhaps as many as half of all spirals have a distinct bar of stars running through the nucleus. They are known as barred spirals, and the spiral arms curve out from the ends of the bar.

In the Hubble classification, spirals are denoted S and barred spirals SB. The letter a, b or c follows, indicating the openness of the spiral arms: an Sa galaxy has the most closed arms, an Sc the most open. Our own Galaxy is traditionally classified as an Sb, since it has moderately open arms, though with evidence of a slight bar through the nucleus, it is sometimes classed as an SBb.

Irregular Galaxies

Irregulars, with no particular shape, are not all that common. In general, they are comparatively small and contain relatively few stars. Yet they are rich in clouds of gas and dust, in which stars are being born. The two Magellanic Clouds visible to the naked eye in far southern skies are irregulars (see page 76).

Hyperactive
The otherwise undistinguished southern constellation of Circinus hosts this unusual looking galaxy. It is a kind of active galaxy known as a Seyfert, noted for its particularly bright center, which seems to be ejecting gas into the surrounding space.

below

Starburst ring
Quite close, at a distance of 30 million light-years, this magnificent galaxy is NGC 1512. It is distinguished by a remarkable outer ring that is a cauldron of star formation. It is a barred-spiral galaxy, although the bar isn't actually visible in this image.

above

Edge-on
If we could view our own Galaxy edge-on from far out in space, it would look like this. This galaxy is NGC 4013 in Ursa Major. The dark band that cuts it in two is thick dust blotting out the light from background stars.

following pages

Galactic fireworks
Incandescent gas clouds and hot star clusters show that vigorous star formation is taking place in this nearby galaxy (NGC 4214). The central cluster is made up of hundreds of blue-white stars, each over 10,000 times brighter than the Sun.

73

Magellan's Clouds

The early sixteenth century was an exciting time to be alive if you were of an adventurous disposition. Italian-born Christopher Columbus had recently sailed from Spain across the Atlantic and discovered a New World, in 1492. Vasco da Gama had sailed from Portugal round the Cape of Good Hope to India , in 1498. This voyage fired the imagination of his young fellow countryman Ferdinand Magellan, then just 18 years old.

Twenty-one years later, now an experienced seaman and navigator, Magellan commanded an expedition that would become the first to circumnavigate the globe. Alas, he never completed the voyage, because he was killed in a skirmish with tribesmen in the Philippines in 1521.

Navigating for months under southern skies, Magellan would have become familiar with the dazzling southern constellations, particularly Crux (the Southern Cross). He would also have noticed the two misty patches visible nearby, and these were since named after him. They are the Large and Small Clouds of Magellan, or Magellanic Clouds.

Also named Nubecula Major and Minor, the Magellanic Clouds look like nebulas, but aren't. They are neighboring galaxies—two of only three galaxies we can see with the naked eye. The other is Andromeda (see page 102).

The Large Magellanic Cloud (LMC) is closest to us, at a distance of about 160,000 light-years. It contains much the same mix of stars and gas as our own Galaxy and has some structure. But at only about 30,000 light-years across, it is not big enough to develop into a spiral galaxy such as our own. One striking feature of the LMC is the Tarantula Nebula, named for its spidery appearance. It is one of the biggest and brightest nebulas we know and is easily visible through binoculars. The Small Magellanic Cloud (SMC) is only two-thirds as big across as the LMC and is about 30,000 light-years farther away.

The Magellanic Clouds are not only close neighbors in space, they are actually companions—satellites—of our own Galaxy. They circle our Galaxy once every 1.5 billion years or so, traveling in an elliptical orbit. Every time they make their closest approach, our Galaxy's powerful gravity attracts some of their stars and gas. As a result, the SMC is already showing signs of breaking apart. In time, both galaxies will be absorbed by our own.

right and far right inset
Heart of the Tarantula
The Tarantula Nebula, also known as 30 Doradus, is the largest stellar nursery in the local universe. Fifteen HST images have been combined to reveal the center of this hotbed of star formation.

below

Inside the LMC

This is just one of the hundreds of star-forming stellar systems in the Large Magellanic Cloud. Unlike ground-based observations, which show the bright blue giant stars, the HST reveals a large number of low-mass infant stars. Inset (*left*) is an infrared image of the heart of the LMC.

Active Galaxies

Galaxies give off energy as light, heat and other invisible radiation, such as X-rays and radio waves. Most of them give out the energy you would expect from a collection of billions of stars. But just a few—about one in 10—give off exceptional energy, maybe millions of times greater than you might expect. They seem to pump out this energy from a tiny region at their center not much bigger than the solar system. Astronomers call these galactic mavericks active galaxies.

In 1943, US astronomer Carl Seyfert noted that some spiral galaxies have exceptionally bright centers. We now know that these Seyfert galaxies are one kind of active galaxy. Since then, several other kinds of active galaxies have been discovered, including radio galaxies, quasars and blazars.

Seyfert Galaxies

These active galaxies are among the most obvious because they mainly emit their exceptional energy at visible wavelengths. Analysis of their spectrum indicates that they contain clouds of hydrogen gas swirling at very high speeds around the galactic center.

Radio Galaxies

At radio wavelengths, these active galaxies appear among the biggest objects in the sky. The radiation seems to come not from the galaxy itself but from regions—radio lobes—on

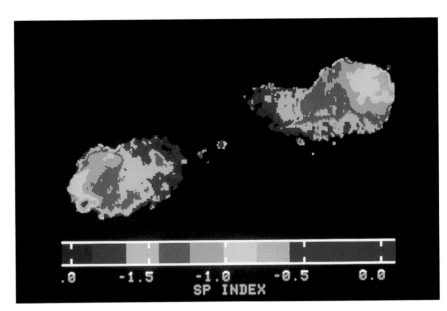

above
Swan song
The second most powerful radio source in the heavens is located in Cygnus (the Swan). Called Cygnus A, it is an active galaxy that pumps out a million times more energy at radio wavelengths than our own Galaxy. The energy is broadcast mainly from two lobes, each 200,000 light-years from the galaxy's center.

below
Along the jets
Combined HST and radio images of the radio galaxy 3C-368 show a string of bright knots that may be stars or pockets of dust. This suggests that the jets streaming from black holes at the center of active galaxies might trigger star formation along their path.

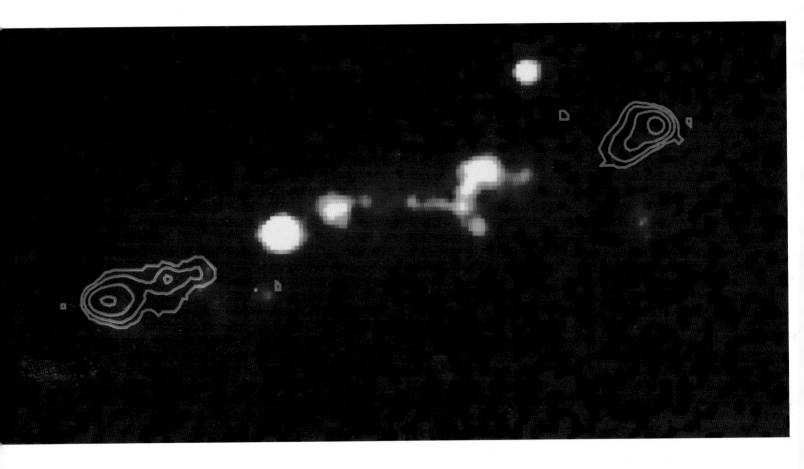

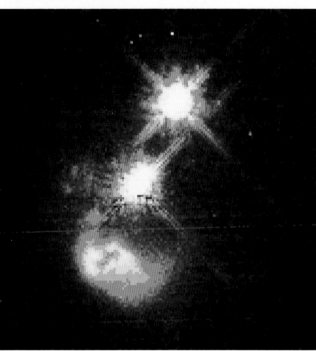

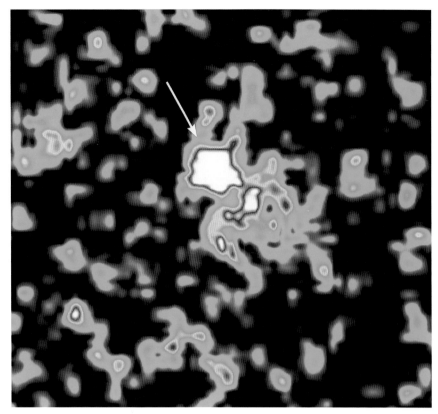

either side and very far out. Although the galaxy may measure only 100,000 light-years across, from lobe to lobe the radio source may span millions of light-years.

Quasars

In the third Cambridge catalogue of radio sources is one in the constellation Virgo, numbered 273, so it is identified as 3C-273. In 1963, when the Moon passed in front of this source and blotted out its signals, astronomers identified it with a star and took its spectrum.

When US astronomer Maarten Schmidt checked the spectrum, he found that it was like that of no other star he had ever seen. This "star" had a huge red shift of spectral lines that placed it at a distance of more than two billion light-years!

3C-273 was clearly no ordinary star, but quite a different body altogether. To be visible at such a distance, it had to be hundreds of times brighter than ordinary galaxies like the Milky Way. Because it appeared star-like, this newly identified celestial body was named a quasi-stellar radio source, or quasar.

Since that time, thousands of quasars have been identified, giving off energy at X-ray and infrared wavelengths as well as light and radio waves. They are all remote, and some are the most distant objects we know in the universe. The quasar PKS 2000-330, for example, is at a distance of 13 billion light-years. This means that we are seeing it as it was close to the time when we believe the universe was born.

above
Gamma-ray burster
The HST spots the optical counterpart of an intense gamma-ray burst (GRB 970228) in the outer reaches of a remote galaxy. Astronomers reckon that the collision between two neutron stars might have triggered the violent release of energy.

left
Quasar host galaxies
The HST has targeted many quasars, revealing details of the host galaxies. Some galaxies seem remarkably undisturbed by quasar activity. In others, the quasars seem to be fuelled by the debris from collisions between galaxies.

Spiral jet

The HST's Advanced Camera for Surveys (ACS) has spotted a giant radio jet coming from a spiral galaxy (0313-192). Here, the ACS image is shown superimposed on a radio image from the Very Large Array radio telescope. It is the first time that a radio jet has been spotted coming from a spiral galaxy.

below

Blowing bubbles

A black hole lurks at the center of NGC 4438, a member of the Virgo cluster of galaxies. Here, the HST has spied glowing bubbles of hot gas rising from the accretion disk surrounding the black hole.

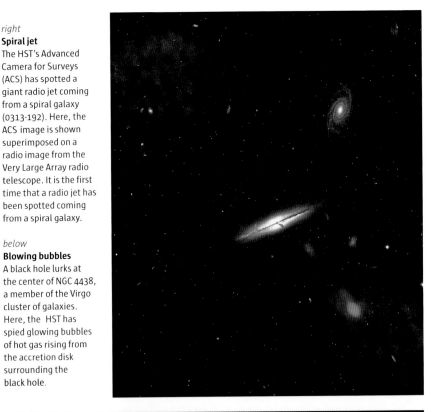

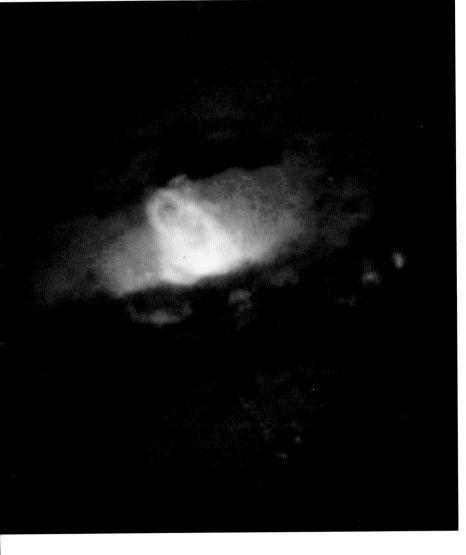

Varying in Brightness

Quasars appear to be a very compact source of energy, much smaller than ordinary galaxies. But just how big are they? We can find out because of how their brightness varies. The brightness of a quasar can fluctuate on a timescale of a day or less.

But it is a fact that the shortest time in which an object can vary in brightness is the time it takes for light to travel across it. This means that a quasar that varies in brightness in a day must be only about a light-day across. This makes it comparable in size to our solar system.

Blazars

A blazar is another kind of active galaxy that varies widely at optical wavelengths. The name comes from a contraction of the terms BL Lacertae and quasar. BL Lacertae is an object first classified as a variable star but later found to be a strong radio source. It is notable because it does not show any lines in its spectrum.

The Energy Machine

Clearly, for active galaxies to be able to pump out such prodigious energy, they must have a powerful "engine." Astronomers reckon that the only engine powerful enough is a supermassive black hole.

Black holes are so-called because they have such powerful gravity that nothing can escape from them — not even light. Relatively small black holes are created when big stars die (see page 62), but supermassive black holes are formed by the collapse of matter in the center of youthful galaxies.

Black holes produce energy when matter is sucked into them. But matter does not simply travel straight down. Because of a galaxy's rotation, matter is strung out into an accretion disk. Surrounding the disk is a donut-shaped torus (ring) of cooler gas and dust.

The accretion disk rotates rapidly and gets so hot that it emits X-rays and other forms of electromagnetic radiation. The radiation can't escape through the disk, and is instead beamed along the rotating axis. Subatomic particles produced in the whirling disk also pour out along the axis, forming high-speed jets. The jets give off radio waves as they encounter particles in the surrounding space.

Point of View

Astronomers believe that the various kinds of active galaxies are different views of the same black-hole structure. When our line of sight is along the plane of the disk, the cool torus obscures the bright center. But we can detect the radio-emitting regions on each side, created by the jets smashing into the intergalactic medium, and so we see a radio galaxy. When the disk is angled more toward us, we see a quasar or a Seyfert galaxy. And if the disk appears face-on, with a jet pointing directly at us, we see a blazar.

above
Sensational Seyfert
A stunning face-on view of the Seyfert galaxy NGC 7742. A massive black hole lurks in the very bright center of this active galaxy.

PROFILING:

Centaurus A

With a visual magnitude of 7, Galaxy NGC 5128 is one of the brightest galaxies in the heavens, only just beyond naked-eye visibility. It is not included in Charles Messier's catalogue of clusters and nebulas and, therefore, does not have an M number.

This was not an oversight on Messier's part, but a reflection on the galaxy's location deep in southern skies and beyond Messier's reach. It lies in the constellation Centaurus (the Centaur) just north of Crux (the Southern Cross).

English astronomer John Herschel (son of William, who discovered Uranus) was one of the first European astronomers to study the southern skies, from South Africa, in the 1830s. He declared the galaxy to be "a most wonderful object, cut asunder by a broad, obscure band."

And NGC 5128 is a strange, beautiful object when viewed through a telescope. It is spherical in shape and is bisected by dark dust lanes. It lies about 15 million light-years away. In 1949, radio astronomers in Australia found a powerful radio source in Centaurus, which they called Centaurus A. They soon identified it with NGC 5128.

Astronomers speculated on the nature of this peculiar bisected galaxy: Could it be a spiral galaxy in the making, and the dark lanes part of a dusty disk that will eventually form spiral arms? Or could its appearance result from a collision between two galaxies — an elliptical and a spiral? Some astronomers favored the latter, reckoning that the radio emission was generated by the collision.

We now know that NGC 5128, usually called Centaurus A, is an active galaxy. It is the nearest active galaxy to us and the third most powerful radio source in the heavens (after Cassiopeia A and Cygnus A). It has been intensively studied by the most powerful radio telescopes and at other wavelengths by the HST, which peers deep into its mysterious dark dust lanes.

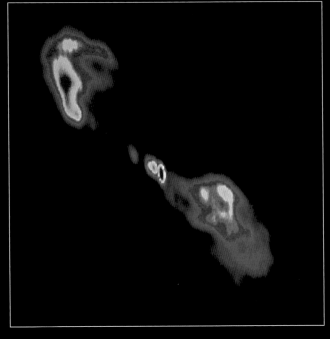

above
On the radio
This radio image of Centaurus A shows the typical wide-spaced lobes of radio galaxies. They extend about 20,000 light-years on either side of the galaxy's visible center.

below
Peering into the dust
In the HST image at left, clusters of hot blue stars line the edges of the dust lane, where star formation is most intense. An infrared view (right) shows a swirling mass of hot gas, caught up in the gravitational whirlpool of a massive black hole.

right
The bright center
In visible light, Centaurus A is one of the most intriguing and unmistakable objects in southern skies. The picture shows the bright central region of the galaxy, measuring about 30,000 light-years across, only about a fifth of its true extent.

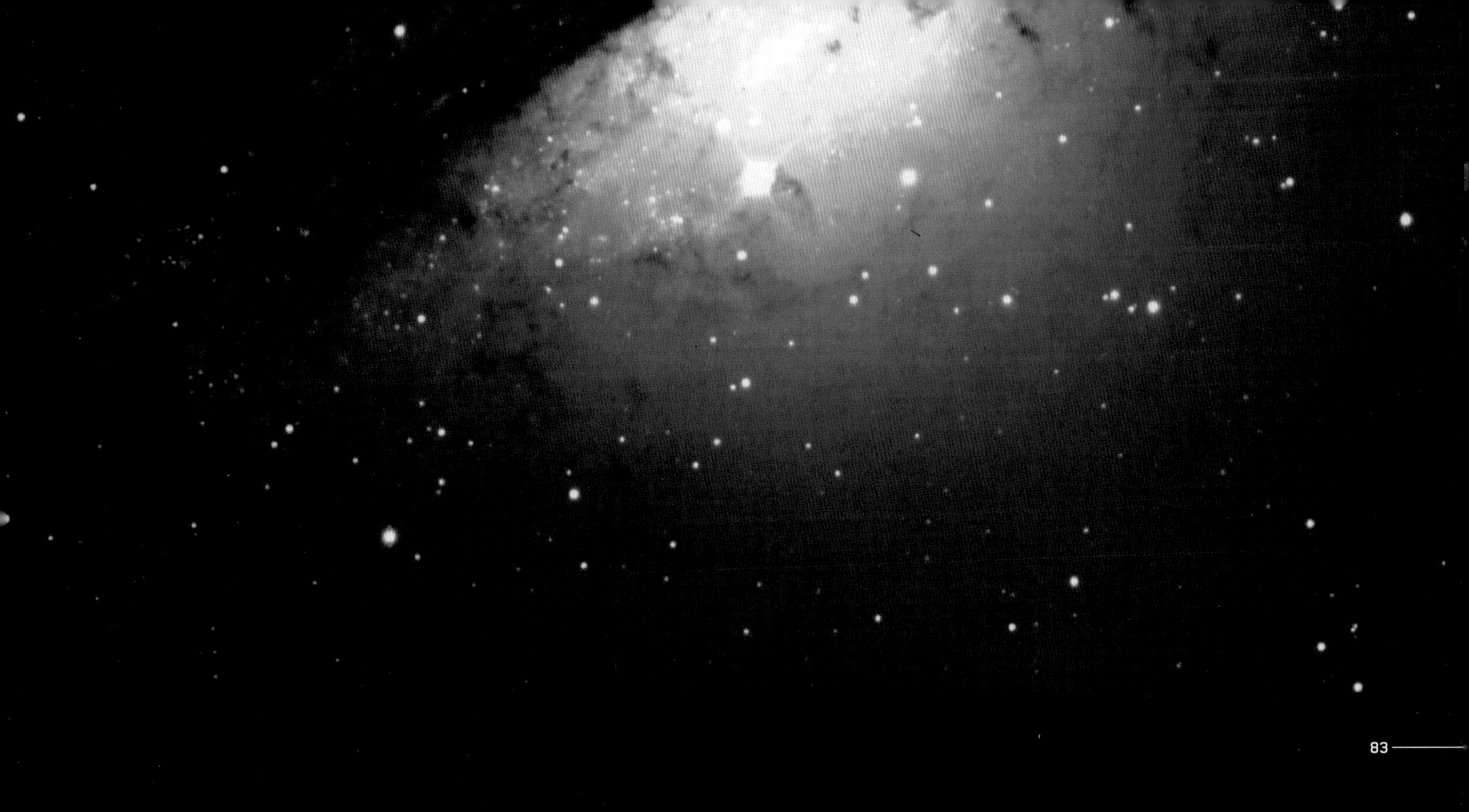

Galaxies in Collision

The strange appearance of Centaurus A is not the result of an intergalactic collision. But collisions between galaxies do occur, and what celestial pyrotechnics they trigger.

Imagine two galaxies, each containing hundreds of billions of stars, closing in on each other at speeds approaching a million miles per hour. They are not solid objects, of course, and don't just bump and bounce apart. Each galaxy's stars begin to infiltrate the other's. The stars don't collide physically, because they are too far apart. Rather they run the gauntlet of a gravitational tug-of-war, being tugged one way by one galaxy, another way by the other. The ultimate outcome of the collision—which galaxy is the victor and which is vanquished—depends on the nature and relative mass of the two galaxies.

Stellar Streamers

If two spiral galaxies of similar mass collide, their stars are pulled from their original orbits, and the galaxies start to lose their spiral structure. Some of the stars gravitate toward the center of mass of the merging star systems. Others are flung out into intergalactic space in long streamers.

A billion years ago a high-speed collision between NGC 4038 and NGC 4039 produced a pair of long, curving streamers of stars. These give the merged galaxies the appearance of an insect's head and antennae, so the interacting pair are called the Antennae.

If the colliding galaxies are unequal in size, several things might happen. As the galaxies draw together, the smaller one could be pulled apart and merge imperceptibly with the larger one. This is happening with our own Galaxy, which is pulling apart its companions, the Magellanic Clouds (see page 76).

Or the small galaxy can also pass right through the large one, destroying its spiral structure and triggering a new round of star formation. This has happened spectacularly in the aptly named Cartwheel Galaxy, with vigorous star formation

above
Cosmic black eye
At first glance M64 appears to be a fairly normal spiral galaxy. But studies have shown its outer regions are rotating in the opposite direction to the inner regions. This unusual galaxy, nicknamed the Black Eye because of the dark dust band in front of its bright nucleus, is the result of two galaxies merging more than a billion years ago.

occurring at the "rim" of the wheel. Astronomers reckon that the event happened around 500 billion years ago.

When the gas clouds in colliding galaxies crash together, or are compressed by gravitational forces, they collapse. This results in a sudden burst of star formation, an event called a starburst. Starbursts also occur in galaxies that are being distorted by larger ones nearby. A good example is M82, which is being disturbed by the larger M81 nearby.

Mergers and Takeovers

Generally, when small and large galaxies interact, the larger one swallows up the smaller and gets even bigger. Over time, astronomers reckon that most galaxies will merge with their neighbors. It seems likely that our own Galaxy, for example, will merge with the Andromeda Galaxy in about five billion years. So, gradually, the nature of the universe will alter as it becomes populated by fewer but larger galaxies.

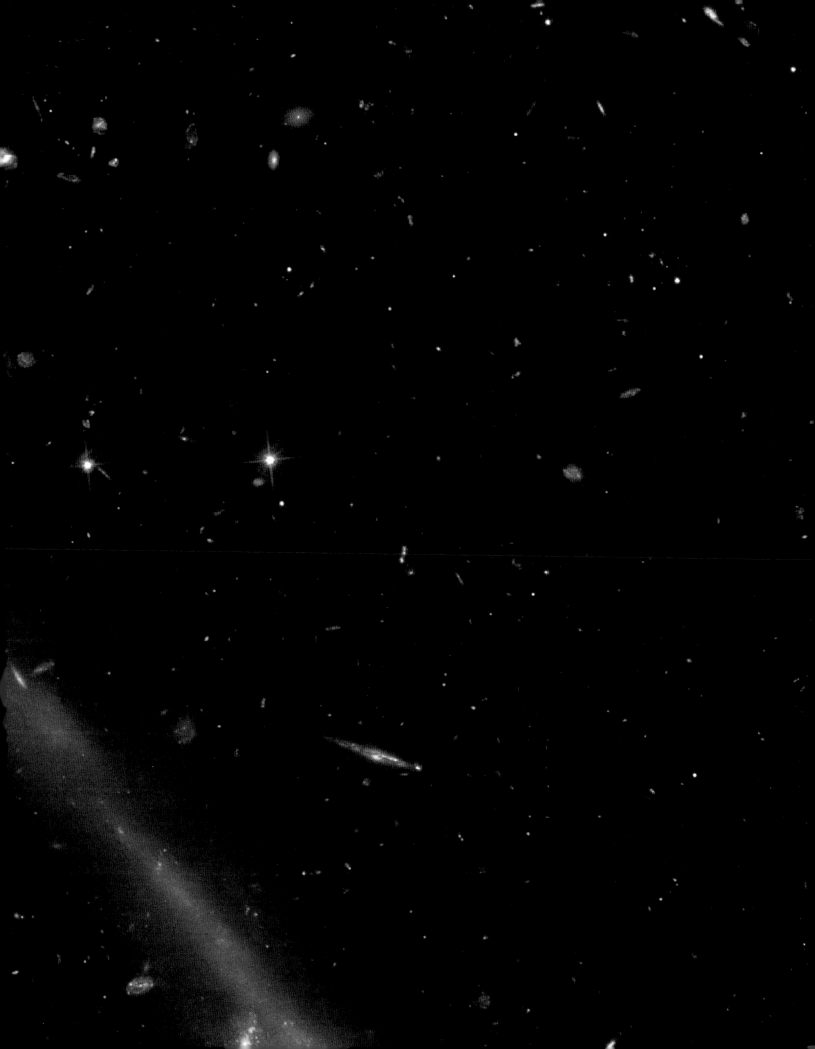

In the Whirlpool

William Parsons, the third Earl of Rosse, was an Irish landowner, whose home was at Birr Castle, Parsonstown, not far from Athlone in central Ireland. Upon graduating from Oxford in 1822, he first dabbled in politics, but abandoned that for astronomy in 1834.

Not content to buy a telescope, Lord Rosse decided to build his own. By 1838, he had built a reflector with a 36-inch (91-cm) mirror. It was an excellent instrument which encouraged him to attempt one twice as large. No one had ever tried to build one this size before. In 1845, he completed the 72-inch (1.8-m) reflector. It was a monster. The mirror was housed in an iron tube nearly 60 feet (18 m) long and 8 feet (2.4 m) across. It was mounted between two massive stone walls and was aptly named the Leviathan of Parsonstown.

With his new telescope, Lord Rosse decided to start observing the star clusters and nebulas that Charles Messier had listed in his catalogue. When he turned the Leviathan on the nebula with the Messier number 51, he was astonished to find that it was shaped like a pinwheel. It had "spiral convolutions," he said. He was the first to discover the spiral nature of the nebulas that would in the 1920s be recognized as external galaxies.

M51 lies about 20 million light-years away in the constellation Canes Venatici (the Hunting Dogs). It is easy to find with powerful binoculars or a small telescope, because it lies just south of Alkaid, the first star in the handle of the Big Dipper. Its nickname, the Whirlpool Galaxy, is apt because we see it face-on, with its spiral arms beautifully revealed.

M51 is actually not one galaxy, but two. The main spiral is NGC 5194, and one of its arms is tenuously linked with a smaller galaxy, NGC 5195. The present structure has resulted from a glancing collision between the two galaxies that probably took place around 300 million years ago. The smaller galaxy was by then an ordinary spiral, or more likely a barred spiral. As it slammed into the outer regions of the large spiral, NGC 5195 was torn apart. Most of its stars were strung out to form a bridge with the other galaxy.

For its size, M51 is particularly bright, with an abundance of young, hot blue stars. Though only about two-thirds as big across as the Milky Way, it is three to four times brighter. Powerful bursts of radio waves come from the centers of both galaxies. The exceptional activity taking place in them almost certainly stems from the collision, as usually happens when galaxies interact.

right
Heart of the Whirlpool
On the HST's 15th anniversary in April 2005 it returned to the celebrated Whirlpool Galaxy. The two curving arms are shown in striking detail. Clusters of young stars are highlighted in red.

below left
Bridging the gap
This wider-angle ground telescope image shows well the Whirlpool's widespread spiral arms and the bridge of gas between it and its companion.

below right
In false color
This image demonstrates the combined results of imaging the Whirlpool Galaxy at different optical and radio wavelengths, shown here in the different colors.

4 | The Expansive Universe

The universe might continue to expand forever

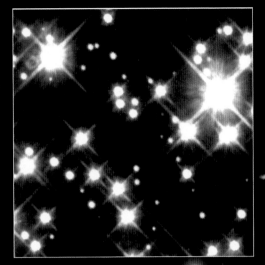

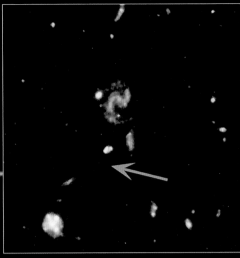

above
Cosmic magnifier
The gravity of a cluster of galaxies has worked like a lens and bent the light from a more distant quasar. Five separate images of the quasar are produced surrounding the cluster's center. The stretched arcs are more distant galaxies lying beyond the cluster.

inset left
Heart of a globular
The HST looks into the heart of the globular cluster M4, which is 5,600 light years away in the constellation of Scorpius. It has identified individual stars, such as white dwarves and pulsars, amongst the 100,000 plus old stars within the cluster.

inset right
Standard candle
With its phenomenally sharp vision, the HST helps pinpoint "standard candles" in remote galaxies. These objects help astronomers make more accurate estimates of the scale of the universe.

After the Big Bang

Everything that exists—our bodies, the air we breathe, the Sun, the stars, the galaxies and space itself—makes up the universe. From year to year we see our local universe, the solar system, change markedly. Changes in the stellar heavens and in the universe at large are much more subtle, but they do occur.

The universe is changing and evolving. But where did it come from? How has it evolved? Exactly how is it made up? And what will happen to it in the future? The astronomers who seek to answer such fundamental questions are known as cosmologists. Cosmology is the study of the origins and evolution of the universe.

Irishman James Ussher, who became archbishop of Armagh in 1625, was no cosmologist, but using biblical references he calculated that the creation of the Earth and the heavens took place on an October morning in the year 4004 BC. He was far off—by a factor of nearly a million. Studies of radioactive rocks suggest that the Earth, along with the rest of the solar system, was born about 4,600 million years ago.

The universe itself is much older. Astronomers believe that it came into being between about 12 and 15 billion years ago. How do they know this? Well, they find that the galaxies are all rushing away from us and from one another. The universe seems to be flying apart, expanding, as though from a massive explosion long ago.

And astronomers reckon that a kind of explosion really did take place—they call it the Big Bang. By measuring the rate at which the universe is expanding now, and calculating backwards, they have determined when the explosion happened—when the universe was created. It was between 12 and 15 billion years ago.

What happened before the Big Bang? This question is impossible to answer because when the Big Bang happened, time itself began. So there can be no "before."

Cosmologists think they know what happened after the Big Bang—except for the first 10^{-43} seconds. During this period, called Planck time, space and time itself, the fundamental forces of nature, and the laws of physics were still forming. After this time, cosmologists can apply knowledge of fundamental forces and laws to describe what happened next.

The universe remained tiny until it was 10^{-35} seconds old. It was incredibly hot (around 10^{27} degrees) and full of energy. Gravity had already become a distinct force, and the first fundamental particles, including electrons, began to form. The universe then expanded suddenly, in an event called inflation. As the universe ballooned, its temperature dropped rapidly.

When the universe was one-millionth (10^{-6}) of a second old, protons and neutrons started to form. Protons are the nuclei (centers) of hydrogen atoms. By about three minutes, the temperature of the universe was only about one billion (10^9) degrees. Protons and neutrons began to stick together, forming the nuclei of deuterium (heavy hydrogen) and helium. At this time the proportions of hydrogen and helium in the universe were fixed.

For about the next 300,000 years, things quietened down somewhat. The universe continued expanding and cooling, but it stayed essentially the same—full of radiation and a mass of atomic nuclei in a sea of electrons. It was a "foggy" universe. Because of all the particles milling around, radiation couldn't travel far without being scattered, like light in a fog

About 300,000 years after the Big Bang, temperatures fell to about 5,000 degrees Fahrenheit (3,000°C). Electrons could now combine with nuclei to form the first atoms of hydrogen and helium. The "fog" cleared as particle numbers were drastically reduced, allowing radiation to pass relatively unhindered. The universe became transparent.

Cosmologists estimate that some time within the first one or two billion years, hydrogen and helium began to form clouds. These would later collapse and create the first stars and galaxies. The HST has made a systematic search for the first bright galaxies to form in the early universe. Hundreds of bright galaxies have been found at around 900 million years after the Big Bang, but very few before then.

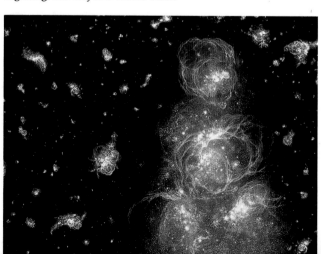

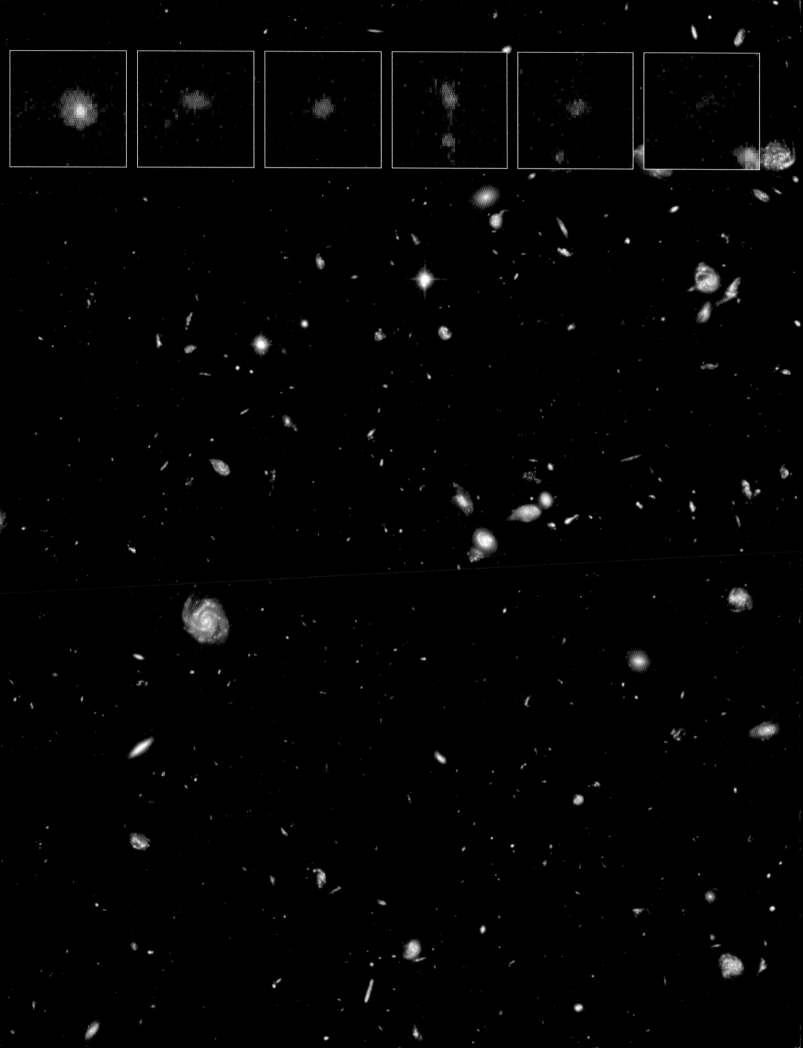

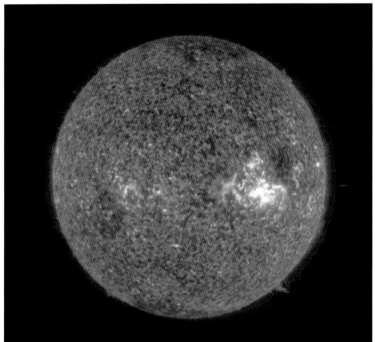

The Scale of the Universe

How big is this universe of ours? Unimaginably vast. But this is a relatively recent realization. Early astronomers knew that the universe was bigger than Earth (which they thought was its center), but they did not believe it was much bigger. Later, astronomers coming to terms with Copernicus's concept of a solar system, equated this solar system with the universe.

Starting in William Herschel's time, in the eighteenth century, people considered the Milky Way Galaxy to be the whole universe. In 1918, US astronomer Harlow Shapley calculated its size to be 300,000 light-years; it turns out this estimate was three times too big. But the true scale of the universe did not became apparent until five years later, when Edwin Hubble proved that spiral "nebulas" were actually separate and remote galaxies, millions of light-years away.

Since then, astronomers have progressively pushed back the boundaries of the universe, using ever-more-powerful telescopes and, more recently, space telescopes like the HST.

From Atoms to Superclusters

The tiniest objects that exist in the universe are particles within atoms — the basic constituents of ordinary matter. Atoms measure about 10^{-10} meters across, or 1/10,000,000,000th of a meter (atomic sizes are expressed as fractions of a meter, never in fractions of an inch). This means that it would take 10 billion atoms, side by side, to measure a meter. A hundred thousand times smaller still, at around 10^{-15} meters across, are the electrons that circle the atomic nucleus. They are among the smallest of all atomic particles, slightly bigger than particles known as quarks (10^{-16} meters).

We need to scale things up a billion times or so to bring us into the realm of our everyday life, when measurements are made in centimeters and meters (or inches, feet, yards, and so on). A few million times bigger is the Earth itself, at about 10^7 meters. And a billion times bigger still is the entire solar system, at about 10^{16} meters (10^{13} km).

When the numbers get this large, measurements expressed in meters or even kilometers become meaningless, so we switch to expressing distances in light-years: one light-year is about 10^{13} km. The solar system is thus about one light-year across.

Multiply the size of the solar system by 100,000, and we get the size of the next-biggest entity, the Milky Way Galaxy. Multiply it 50 times more, and we're up to the scale of the cluster of galaxies to which the Milky Way belongs. Twenty times more, and we're looking at a supercluster. Two-hundred times more, and we're nearing the edge of the observable universe, more than 10 billion light-years away.

This distance works out at 10^{26} meters. The scale of the universe — from the smallest object to the largest, from a quark to the observable edge of space — goes from 10^{-16} to 10^{26} meters. In other words, the universe is more than a million, million, million, million, million, million, million (10^{42}) times bigger than the smallest particle.

above left
Planet Earth
Our verdant home in space, where conditions are ideal for life to thrive in abundance.

above right
The Sun
Our local star, which pours heat, light, and other radiation into space — we see it here through the eyes of the SOHO spacecraft.

right inset top
Stars
The far-distant suns that shine in the night sky, as here in the constellation Cygnus, one of the few that lives up to its name (the Swan).

right inset bottom
A galaxy
A great star island in space, populated by billions of stars.

right (main image)
The deep universe
Galaxies galore appear in deep space images. This typical patch of sky imaged by the HST includes galaxies, far and near, big and small.

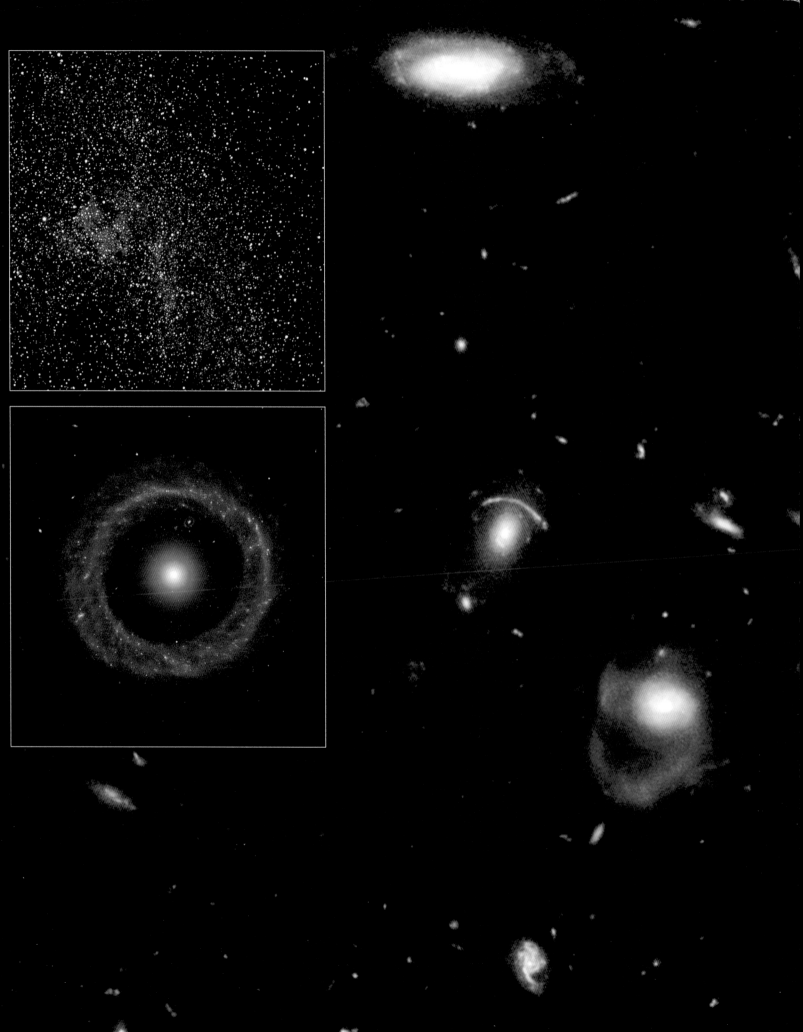

Cosmological Distancing

Proxima Centauri, the nearest star to the Sun, lies about 4.3 light-years away from Earth. Sirius, the brightest star in the heavens, lies 8.8 light-years away. The Andromeda Galaxy, the farthest object we can see with the naked eye, lies 2.5 million light-years away. Some galaxies lie more than 10 billion light-years away.

How do we know that these figures are accurate? How do we measure the distance to the stars and galaxies? To measure the distance to a few hundred of the nearest stars, astronomers use a variation of the technique surveyors use to measure distances on land.

In this process, called triangulation, a surveyor first measures out a baseline—a straight line along the ground— and then measures the angles from each end of the baseline to a distant point, for example, a church. Knowing the length of one side (the baseline) and two angles, the surveyor can use simple trigonometry to calculate the lengths of the other two sides of the triangle formed by the ends of the baseline and the church. This pinpoints the location of the church.

Using Parallax

For the astronomical version of triangulation, astronomers choose a much longer baseline: the diameter of the Earth's orbit around the Sun. They view a star from one side of Earth's orbit in January, for example, and then from the other side six months later. A nearby star seems to change position slightly against the background of distant stars during this time. This

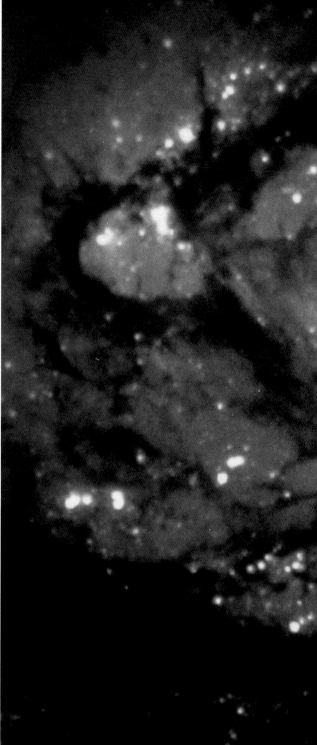

right
M100
The remote spiral galaxy M100, one of thousands in a mammoth cluster of galaxies in Virgo. The HST can pinpoint individual stars in the spiral arms, in particular Cepheid variables, which can be used as standard candles.

below
Hipparchos
Named for the foremost astronomer of ancient Greece, Europe's Hipparchos satellite was launched in 1989. It spent over three years measuring the parallax of some 120,000 stars with unprecedented accuracy.

April 23

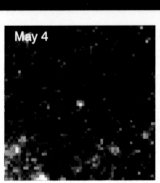
May 4

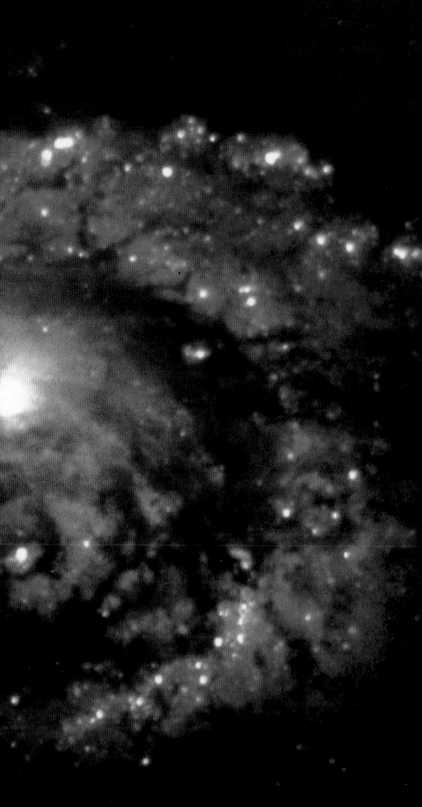

effect is called parallax. You can easily demonstrate parallax yourself. Hold up a finger in front of you and look at it first with one eye and then the other. You'll notice that your finger seems to shift position against the distant background, and the nearer your finger is to your face, the more it seems to shift.

From the parallax shift shown by a star, astronomers calculate its distance in the same way a surveyor would for the church. However, parallax measurement has severe limitations. Parallax shifts are tiny, even for the nearest stars, and for stars farther than about 100 light-years, they are almost impossible to detect. In the early 1990s, however, the European astronomy satellite Hipparchos managed to measure parallaxes of around 120,000 stars out to a distance of about 500 light-years.

Standard Candles

Parallax measurement is certainly not an option for measuring distances to other galaxies. To measure galactic distances, astronomers look for what are called standard candles. These are objects that are always of similar brightness or behave in a predictable way.

The classic standard candles are variable stars called Cepheids. Early last century, US astronomer Henrietta Leavitt discovered a fundamental law about Cepheids: that the period over which they vary in brightness is directly related to their true brightness (absolute magnitude or luminosity).

This relationship, known as the period-luminosity law, makes Cepheids standard candles. When astronomers measure the period of a Cepheid, they know what its true brightness is. They then compare this true brightness with the star's apparent brightness in the sky. Because light fades in proportion to increasing distance, it is then easy to calculate how far away the Cepheid is.

It was Edwin Hubble who first used a Cepheid as a standard candle to measure the distance to the Andromeda Galaxy, proving that it lay beyond our own Galaxy (see page 162). The HST has extended his work, pinpointing Cepheid standard candles in some of the most distant galaxies.

left
Cepheid variations
From HST observations of the light variations of a Cepheid variable in M100 over several months, astronomers find its period and hence its luminosity, or true brightness. Comparing this with its apparent brightness, they calculate that the Cepheid lies 51 million light-years away.

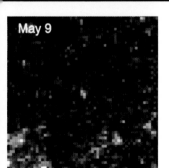
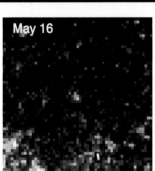
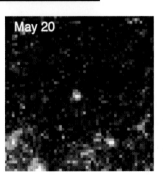
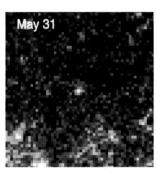

May 9 — May 16 — May 20 — May 31

Other Standard Candles

Globular clusters are other objects that can be used as standard candles. These globe-shaped masses of stars orbit the centers of galaxies, and the brightest ones have about the same true brightness. By comparing this with their apparent brightness in the sky, astronomers can estimate their distance.

Supernovas are useful as standard candles for even more remote galaxies. These exploding stars can become for a while as brilliant as an entire galaxy and can therefore be spotted at enormous distances. The most useful kind of supernova is the Type I, caused by the explosion of a white dwarf star (see page 57), because they all reach close to the same level of brightness.

Red Shift

The most remote galaxies are too far away to detect any kind of standard candle. To determine how far away they are, astronomers revert to the invaluable astronomical technique of spectroscopy.

Just as astronomers can tell a lot about a star by examining the spectrum of its light (see page 32), they can also tell a lot about a galaxy from its spectrum. They see in the spectrum the characteristic dark absorption lines, which identify certain chemical elements. They can tell by the position of the lines in the spectrum whether the galaxy is moving toward or away from us.

below
Standard globulars
Globular clusters can make useful standard candles to measure distances to some galaxies. Here in the core of the giant elliptical galaxy NGC 1275 in Perseus, the HST resolves a number of globular clusters (blue). These are unusual because they consist of young stars, not old.

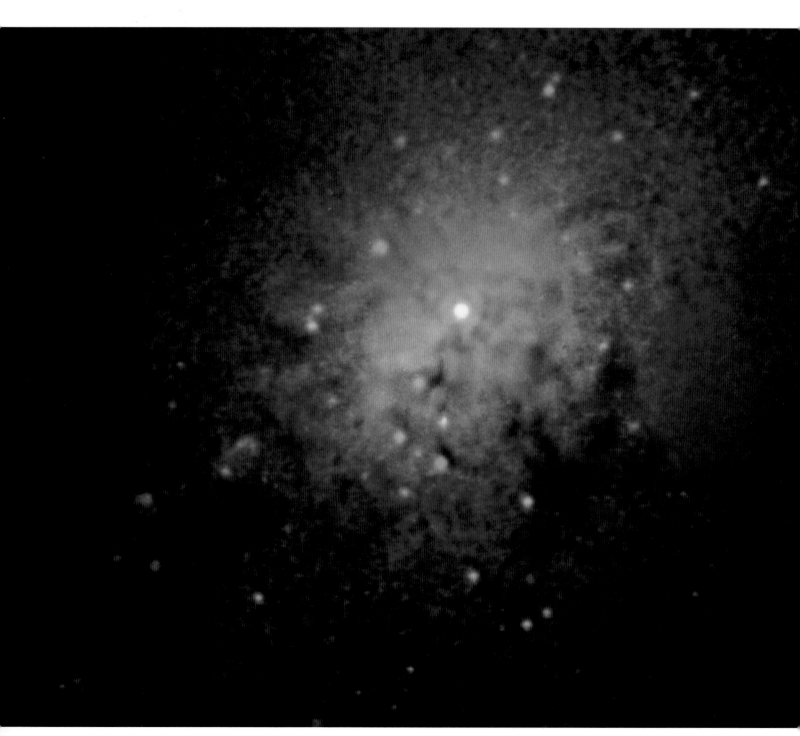

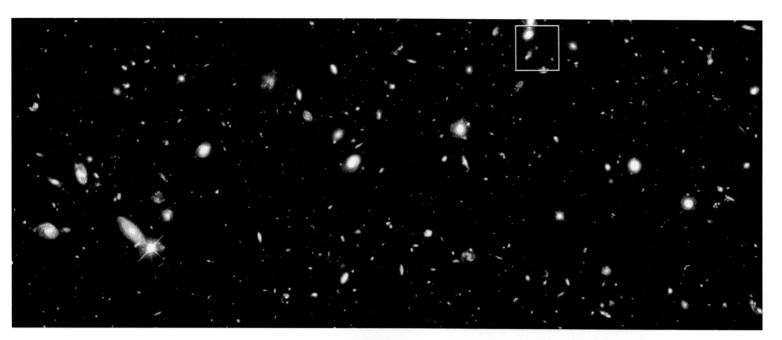

When a galaxy is moving toward us, its light waves bunch up and seem to have a shorter wavelength, which makes them appear bluer. And the galaxy shows a spectrum with lines shifted toward the blue end. This is called blue shift. When a galaxy is moving away, its light waves stretch out and seem to have a longer wavelength, which makes them appear redder. And the lines in the spectrum shift toward the red end. This is called red shift. The extent of the blue or red shift tells us how fast the galaxy is moving.

This bunching up and stretching out of waves from a moving source is called the Doppler effect. We experience this with sound waves when an ambulance — siren blaring — races first toward us and then away. The noise of the siren has a higher pitch (shorter wavelength) when the ambulance is approaching, because the sound waves are bunching up. The noise has a lower pitch (longer wavelength) when the ambulance is going away, because the sound waves are being stretched out.

Relating Red Shift

Astronomers find that almost all the galaxies have a red shift, indicating that they are rushing away from us and from one another. This is how Hubble and other astronomers in the early twentieth century realized that the universe must be expanding.

Hubble then made a discovery that is crucial to cosmological distance measurement. He found that the speed of recession of a galaxy (measured by its red shift) is directly related to its distance. This relationship is known as Hubble's Law. The rate at which galaxies speed up with increasing distance is called the Hubble Constant. Current calculations seem to show that a galaxy speeds up by around 45 miles (70 km) a second every megaparsec (about three million light-years).

By more accurately measuring the distance to remote galaxies, the HST has been helping refine the value of the Hubble Constant. The more accurate the value, the more accurately we can describe and comprehend the universe.

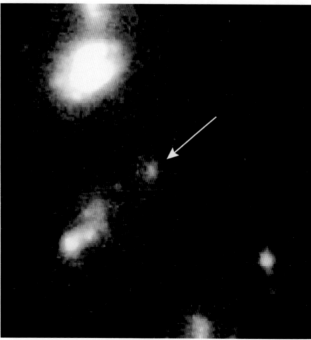

above and left
Ancient blast
Comparing the light output of galaxies in the Hubble Deep Field over an interval of two years, HST scientists discovered in one of them a supernova. The explosion had taken place 10 billion years ago. Supernovas prove useful in estimating distances to remote galaxies.

left
Spectral shifts
The shift of the dark lines in the spectra of stars and galaxies tells us if they are traveling toward us (blue shift) or away (red shift). Spectral shifts can also detect rotations. Here, the HST's spectrograph has scanned the rotating disk around a black hole in the galaxy M84. The colors show the abrupt shift in wavelengths from one side of the disk to the other.

Grouping Together

Galaxies are not distributed evenly throughout space; they gather together into small groups or larger clusters. It is the mutual gravity of galaxies that helps keep them together and overcomes, on a small scale, their tendency to fly apart as the universe expands.

In our corner of space, gravity draws together our Galaxy (the Milky Way) and about 30 other galaxies to form what is called the Local Group. The Milky Way is one of the two large spirals that dominates the Local Group. The other is the Andromeda Galaxy. There is just one other smaller spiral, M33. All the rest are small, dwarf elliptical or irregular galaxies. Altogether, the Local Group occupies a region of space around five million light-years across.

Satellite Galaxies

Within the Local Group, smaller galaxies congregate around the two large spirals. We call them satellite galaxies. The two Magellanic Clouds we can see with the naked eye in southern skies are satellites of the Milky Way. Our Galaxy also has many other satellites, all of them dwarfs and all too far away to see with the naked eye.

The Sagittarius dwarf galaxy, about half as wide across as the Large Magellanic Cloud (LMC), lies even closer to us, at a distance of around 78,000 light-years. This is so close that the gravity of the Milky Way is ripping it apart. Obscured by the dust in the plane of the Milky Way, this galaxy was not discovered until late last century.

Since the Sagittarius dwarf galaxy is nearly spherical, it is classified as elliptical. Many of the Milky Way's other satellites are dwarf ellipticals. They include the diminutive Draco and Carina systems, both around 250,000 light-years away and both only about 500 light-years across — only 1/200th the size of our own Galaxy. They may contain as few as 300,000 stars.

The Other Spirals

The Andromeda Galaxy is a spiral, just like the Milky Way but considerably larger. This magnificent galaxy has figured prominently in the history of astronomy and is profiled overleaf.

The Andromeda Galaxy also has satellites orbiting round it. The third spiral in the Local Group, M33, seems isolated, although many astronomers consider it to be a satellite too.

M33 is often called Triangulum because it lies within the tiny northern constellation of that name, whose three brightest stars form the shape of a triangle. M33 looks magnificent from Earth because we see it face-on and have an excellent view of its spiral arms. With a visual magnitude of just under six, M33 is right at the limit of naked-eye visibility. Observers with exceptional eyesight might spot it under very clear conditions. However, it is easy to see with binoculars, visible in the same field of view as the constellation's lead star, Alpha. One striking feature of the galaxy is a giant nebula of hydrogen gas, NGC 604, which is a vast star-forming region on one of M33's spiral arms.

above
Wide-open arms
The wide-open arms of M33 are well seen in telescopes. The galaxy lies 2.7 million light-years away, a little farther away than the Andromeda spiral. With a diameter of some 40 million light-years, it is less than half the size of our own Galaxy.

right
Star-birth spectacular
The rich populations of infant stars found in N90, one of the star-forming regions in the Small Magellanic Cloud, allow astronomers to examine the process of star formation in an environment outside the Milky Way. Bright blue newly formed stars are blowing a cavity in the center of the region.

The Great Spiral

The constellation Pegasus, the Flying Horse, is a prominent feature of fall skies in the Northern Hemisphere (and spring skies in the Southern). It is easy to recognize because it contains an almost perfect square, the Square of Pegasus.

The star that marks the top-left corner (bottom-right in the Southern Hemisphere) actually belongs to the linked constellation Andromeda. If you cast your eyes up and slightly to the left (down and to the right in the Southern Hemisphere), you will see a faint, misty patch. It looks like a nebula, but isn't—it's a neighboring galaxy.

The Andromeda Galaxy is nowhere near as close as our other galactic neighbors, the Magellanic Clouds. It lies around 2.5 million light-years away and is the most distant object that is visible to the naked eye.

The Arabian astronomer Al-Sufi mentioned the galaxy first in his *Book of the Fixed Stars* in the year AD 964. German astronomer Simon Marius spotted it again in a telescope in 1612, saying that it looked like "the light of a candle seen through horn." Charles Messier included it as number 31 in his list of star clusters and nebulas of 1781.

It seems astonishing that we can see the Andromeda Galaxy, M31, at a distance of millions of light-years away. This is because of its enormous size—150,000 light-years across, which is one-and-a-half times as big as the Milky Way. It also contains many more stars—infrared observations made in 2006 suggest it has about one trillion. It dominates the Local Group and could, in billions of years, consume most of the other galaxies, including the Milky Way.

M31 is a spiral galaxy, also called the Andromeda Nebula and the Great Spiral. Unfortunately, we don't have a very good view of its spiral arms because we see it from the side. M31 was one of the spiral nebulas investigated in the early twentieth century by Edwin Hubble and other astronomers. It was the first one he proved was extragalactic—beyond our own Galaxy (see page 162).

Later studies of the spiral arms and central bulge by Walter Baade (1944) led to his recognition of the two populations of stars. Population I stars are the relatively young stars of spiral arms, while Population II stars are the older stars of the galactic center.

The Andromeda Galaxy is like our own spiral in another way; it has two close companions, the galaxies M32 and NGC 205. They are both dwarf elliptical galaxies in orbit around the Great Spiral.

above
Stars with the blues
Deep inside the elliptical satellite galaxy M32, the HST has spotted a swarm of hot blue stars, seen here in ultraviolet light. Unusually, these stars seem to be old, at a late stage in their evolution when they are burning helium in their cores.

below
Andromeda's globulars
The HST views globular clusters orbiting M31's nucleus with almost the same clarity as ground-based telescopes view globulars in our own Galaxy. This Andromeda globular boasts plenty of red giants, as would be expected. The two other bright objects are nearby stars in our Galaxy.

right
The spiral giant
Our close galactic neighbor, the Great Spiral in Andromeda (M31), is the most famous of all the outer galaxies. It appears to have two major spiral arms. Seen here also are the two satellite galaxies—M32 close in and NGC 205 further out.

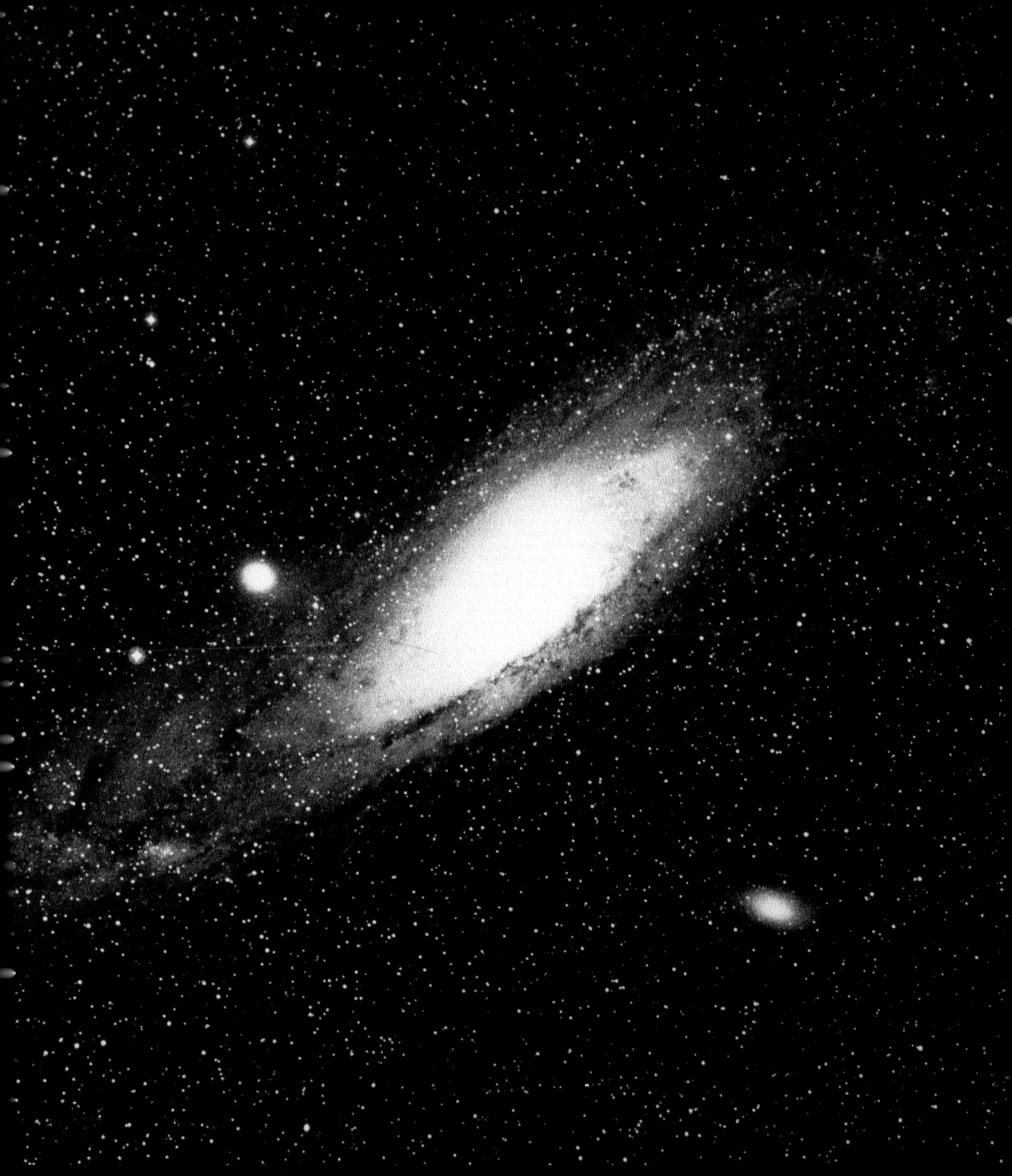

below
Galaxy cluster
The diverse nature of galaxies within a galaxy cluster is revealed in this HST image of Abell S0740. In the center is a giant elliptical galaxy, which is as massive as 500 billion stars. Other fuzzy ellipticals and several spirals are also present.

far right
Stephan's quintet
Four of the galaxies in Stephan's Quintet, a group discovered by Edouard Stephan in 1877. The galaxy near the center and the ones near the top of the image are physically associated. Interactions between them have initiated bouts of star formation and long gaseous streamers. The galaxy at the right-hand edge is much closer to us.

right
Dance of destruction
This group of galaxies is named Seyfert's Sextet for its discoverer Carl Seyfert. But only four galaxies are physically associated and interacting. The small face-on spiral lies much farther away than the others, and the object at lower right is an elongated tail of stars rather than a galaxy.

Clusters and Superclusters

The Local Group is a very small cluster of galaxies. Many small clusters make up a large cluster, and many of these large clusters go into forming superclusters. They extend throughout the universe in sheet- and streamer-shaped structures over distances of more than 100 million light years. The Sculptor Supercluster is about a billion light years away and measures over 250 million light years from end to end.

Clusters of galaxies vary not only in size but also in shape. The Local Group and the Virgo Cluster are irregular in shape, but the Coma Cluster is roughly spherical.

In general, irregular clusters are made up mainly of spiral galaxies, while elliptical galaxies dominate in spherical clusters. Irregular galaxies also seem to be much younger than spherical ones. Astronomers think that this probably reflects how galaxies evolve. Over time, the spirals in irregular clusters tend to collide and merge to form ellipticals, while the cluster itself assumes a more regular shape.

Galactic collisions account for the presence in some large clusters of giant elliptical galaxies. There are three at the center of the Virgo Cluster. Giant ellipticals are worthy of their name because they can measure as much as two million light-years across — about half the size of the entire Local Group.

Superclusters

It is gravity, of course, that holds the galaxies together in clusters. Gravity also binds clusters together to form mammoth superclusters. Our Local Group belongs to a supercluster centered on the big Virgo cluster. Called the Virgo, or Local Supercluster, it occupies a region of space about 100 light-years across.

Superclusters are the biggest structures in the universe. So far more than 50 have been discovered. They may take a variety of shapes, from flat sheets (called filaments) to curving streamers. The largest one we know is a filament called the Great Wall. It measures about 750 million by 250 million light-years.

Like stars and galaxies, superclusters are not distributed evenly throughout space. They seem to be wrapped around empty spherical regions called voids. The voids connect with one another, giving the universe a spongy structure.

This irregular scattering of galaxies and clusters could well reflect the original "lumpiness" of matter after the Big Bang, as inferred from measurements of cosmic background radiation by the COBE satellite (see page 156).

left
Zooming in
Some of the arcs
produced by the
gravitational lensing
effect of the huge
galaxy cluster Abell
1689, revealed in
unprecedented detail
by the HST's Advanced
Camera for Surveys
(see pages 178/179).

A Matter of Gravity

left
Copycats
A superb example of gravitational lensing, created by a cluster of elliptical and spiral galaxies known as 0024+1654. The cluster's lensing effect produces a pattern of arcs (blue). The similar shape of the arcs and their similar red shifts shown that they are views of the same object, a spiral galaxy about 10 billion light-years away.

Some of the most interesting HST images show a phenomenon known as gravitational lensing. Some pictures of distant clusters of galaxies reveal little curved arcs, which appear to be separate, distorted galaxies.

The red shifts of the light from these arcs place them far beyond the cluster itself, and the red shift of all the arcs is the same. It turns out that what we are seeing are repeated, distorted images of the same galaxy.

To find out why this is happening, we must refer back to Albert Einstein's theories of relativity. In his General Theory, published in 1916, he pointed out that gravity can interact with light and bend it. In our example, the cluster has such powerful gravity that it bends the light coming from the distant, hidden galaxy and brings multiple images of it into view. The cluster is acting like a lens—a gravitational lens.

Exactly how a distant galaxy appears as a result of gravitational lensing depends on how the galaxy and the cluster are aligned toward us. If the alignment were perfect, the arcs would merge into a complete, or "Einstein" ring.

The Riddle of Dark Matter

When astronomers calculate how strong the gravity of a cluster must be to produce a lensing effect, they come up against a problem. Without exception, the total mass of the galaxies visible in the cluster is not nearly large enough to produce the powerful gravity needed for lensing. In general, a cluster needs 90 percent more mass to produce this effect. Astronomers reckon that this mass must reside in matter that is invisible. They call it dark matter.

In the Halo

But it is not only in gravitational lenses that dark matter must exist. The way that spiral galaxies rotate and retain their structure over time implies that they are constrained by a gravitational force much greater than their visible matter—their stars and gas clouds—can account for. Each galaxy must be embedded within a vast spherical region, or halo, containing abundant dark matter. The halo, in turn, may be surrounded by an even more extensive sphere, or corona, of dark matter. There could be as much as 10 times more dark matter than visible matter in an average galaxy.

Evidence of dark matter is also provided by galaxy clusters. The galaxies in clusters travel so fast that, in theory, they should rapidly escape from one another. The fact that this doesn't happen suggests that gravity from dark matter must be holding them back.

right
Hot clusters
Part of a cluster containing thousands of galaxies, more than 8 billion light-years away. This is an X-ray image, from a 34-hour exposure by the Rosat astronomy satellite. X-ray images of clusters also reveal pools of very hot gas at temperatures up to 180,000,000°F (100,000,000°C).

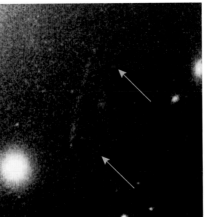

Of MACHOs ...

There is circumstantial evidence, then, that visible matter makes up only about 10 percent of the universe, and dark matter 90 percent. But exactly what is dark matter? Because we can't see it or detect it, we don't know for certain. But it could take a variety of forms.

Ordinary visible matter consists of atoms made up of lightweight electrons and a nucleus of heavier particles, protons and neutrons. These heavier particles belong to a class of subatomic particles called baryons; so ordinary matter is called baryonic matter.

The dark matter in the halos of galaxies could be invisible baryonic matter, such as the failed stars we call brown dwarfs, dead stars and black holes. Such invisible objects are called MACHOs (massive compact halo objects).

... and WIMPs

Astronomers also believe that the universe might be flooded with unseen and so-far undetected non-baryonic forms of matter. Physicists have suggested that this matter could consist of WIMPs (weakly interacting massive particles). WIMPs would be difficult to detect because they would barelyly interact with ordinary matter. Astronomers and physicists are searching for them using specialist telescopes and other detectors.

Other elusive particles that we know exist are neutrinos. Neutrinos produced in the Sun's interior are racing through your body right now. They are elusive because they hardly interact with matter and have no electric charge. Until recently physicists also thought that they had no mass. Current research suggests that neutrinos might have a very small mass—around 1/100,000th the mass of an electron. If this is true, then neutrinos could account for a large proportion of the dark matter in the universe.

Open or Closed?

The amount of dark matter is crucial in deciding the eventual fate of the universe. If there is enough matter, visible and invisible, then its collective gravity will be strong enough to halt the universe's expansion. Otherwise, gravity will not be strong enough and the universe will continue to expand.

Cosmologists have worked out a critical density for the universe at which gravity will be just strong enough to stop it from expanding. It works out to be a few hydrogen atoms per cubic yard (cubic meter). The ratio of the actual density of the universe to the critical density is designated Ω (omega).

There seem to be three main possible fates for the universe. If Ω equals 1, then the universe will eventually stop expanding, but only after an infinite amount of time, and it will exist forever. This concept is called the flat universe.

If Ω is more than 1, then the universe will eventually slow down and stop expanding. Gravity will then rein in the galaxies and the universe will contract. Over time, everything in the universe will come together in a Big Crunch, a reversal of the Big Bang. This concept is called the closed universe.

If Ω is less than 1, the expansion of the universe that we witness today will continue forever. All the stars, all the galaxies will fade and die; all the black holes will evaporate away. Unimaginably cold, unimaginably vast, the universe will end up as an ocean of subatomic particles with no energy or ability to interact. This concept is called the open universe.

So is the universe flat, closed or open? Will it exist forever, will it end in a Big Crunch, or will it fade away? The jury is still out, but the signs indicate that ours is an open universe. And recent observations suggest that the rate of expansion is actually increasing. This has led to the suggestion that a mysterious force called dark energy must be stretching the universe.

5 | Solar Systems

Other stars have planetary systems too

above
Stellar beginnings
Dark dust clouds and hot young stars make up this star-forming region, N11B, in the Large Magellanic Cloud, a nearby galaxy. The blue and white stars to the left are among the most massive stars known in the universe.

inset left
Comet Tempel 1
This HST image of Comet Tempel 1 was taken just days before the Deep Impact spacecraft made its rendezvous with the comet. The comet's inner coma of dust and gas surrounds the cometary snowball nucleus.

inset right
Asteroid Ida
The probe Galileo sent back this image of the asteroid Ida while it was on its way to Jupiter in 1993. Ida measures about 35 miles (55 km) long and, incredibly, has a minuscule moon orbiting around it, named Dactyl.

Planets in the Making

Vast, dark tenuous masses of gas and dust called giant molecular clouds are found throughout interstellar space. They are the stuff that stars are made of, are born from. Five billion years ago, a giant molecular cloud occupied our little corner of the universe. It was from this that the Sun and the planets were born.

Part of the cloud began to collapse under gravity and rotate. As the collapse continued and rotation became faster, the cloud flattened out to form an embryonic solar system that astronomers call the solar nebula. The denser center of this rotating mass began to heat up and glow, becoming the proto-Sun.

Matter continued to feed into the collapsing proto-Sun, which in turn continued to heat up. Eventually, soaring temperatures and pressures triggered nuclear reactions within its core, and it began to shine as a new star. This fledgling Sun then spent roughly 10 million years calming down, until it reached equilibrium between the outward pressure of radiation and the inward force of gravity. It became a stable main-sequence star.

Meanwhile, the surrounding matter had flattened itself into a disk, and it was from this disk that the planets and other members of the solar system eventually emerged.

Scenarios similar to the birth of our solar system are being enacted throughout interstellar space, giving rise to planetary systems around other stars. The first evidence of possible external solar systems came in 1983, when the infrared astronomy satellite IRAS detected dusty disks around the stars Beta Pictoris and Vega. The HST has since spotted dusty disks around a host of stars—dozens in the Orion Nebula alone. They are termed proplyds, for protoplanetary disks, because they are almost certainly planets in the making.

Within our own solar system, how did the planets we know today form out of the primeval dusty disk that girdled the youthful Sun? Surely the planets of other stars would evolve through a similar process.

Inner Planets

In the flattening disk of matter that formed around the proto-Sun, particles began to clump together. Eventually, these clumps grew into masses several thousand yards (meters) across, known as planetesimals. By this time, these great chunks were in orbit around the newborn Sun.

The temperature of the disk of rotating matter varied widely, from a high of around 3,500 degrees Fahrenheit (2,000°C) closest to the Sun to a low of only about −360 degrees Fahrenheit (−220°C) at the outer edge. This variation in temperature dictated the distribution of molecules in the disk.

In the hot, inner region, metals and rock-forming silicate minerals (which have high melting points) collected. Planetesimals with this composition gradually merged by collision to form larger and larger objects, which eventually became the rocky, terrestrial planets—Mercury, Venus, Earth and Mars.

It took roughly 10 million years for these planets to reach more or less their present size, and about another 100 million years for them to sweep up any remaining planetesimals. By this time, powerful blasts of particles in the solar wind had swept the inner solar system clear of gas and volatile substances like water, ammonia and methane.

right
Possible planet?
A dust disk surrounds the star Beta Pictoris whose light has been blocked out in this HST image. The second less-extensive disk tilted to the first could indicate the presence of a planet. The planet may have formed the second disk by sweeping up material from the main disk.

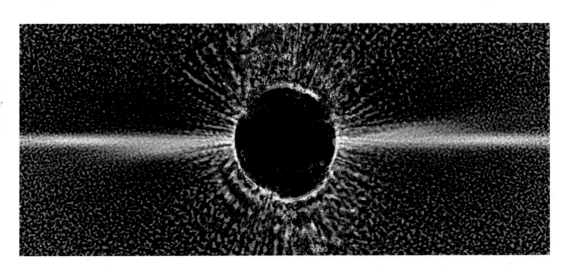

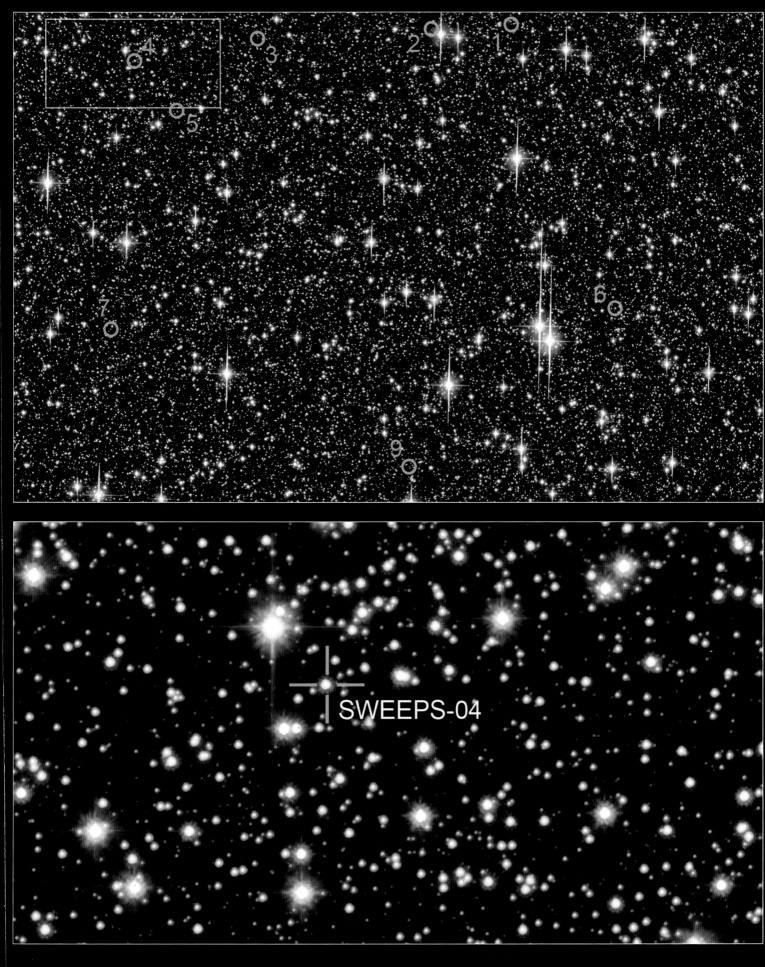

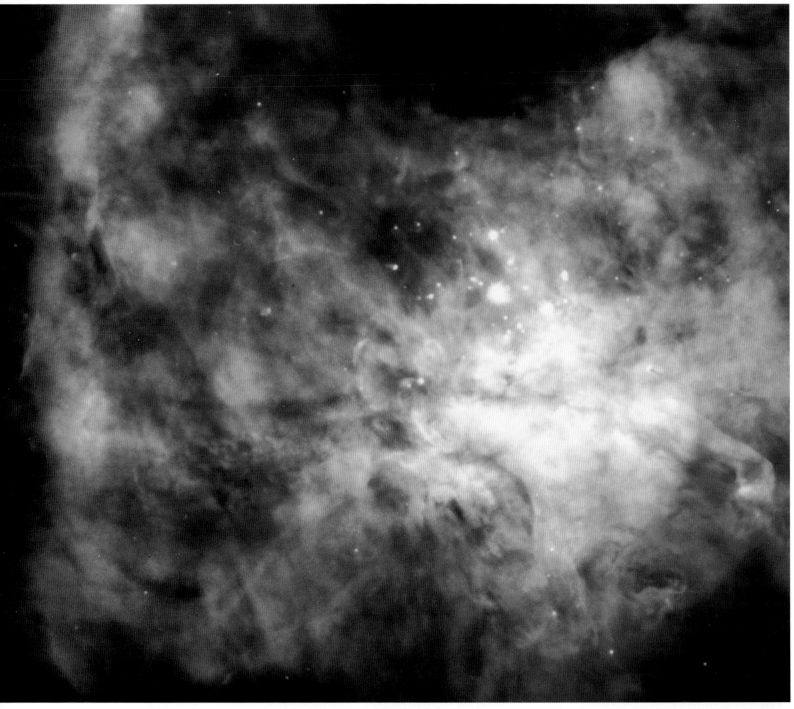

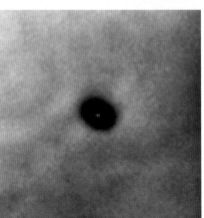
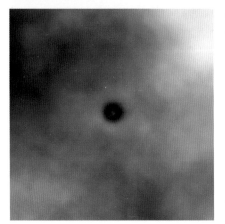
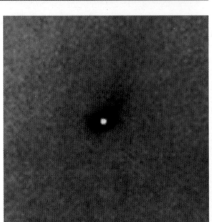

above
Orion proplyds
The HST has imaged more than a hundred protoplanetary disks, or proplyds, in the vast star-forming region that is the Orion Nebula. Clearly, they are a common by-product of star formation.

Outer Planets

Planets were also forming in the outer part of the circumsolar disk, where it was cooler. Hundreds of millions of miles away from the Sun, bodies were growing that would become Jupiter and Saturn. They were growing huge by attracting gas, primarily hydrogen and helium, from their surroundings.

Farther out still, the bodies that would become Uranus and Neptune were created when the volatiles, such as water and ammonia, condensed on small, rocky cores. They accumulated relatively little hydrogen and helium, since Jupiter and Saturn had mopped up most of those gases.

Beyond Neptune, ice was the predominant matter. Relatively small worlds of ice and rock formed, of which Pluto was one of the largest. Beyond these worlds and out to the edge of the solar system, matter remained as icy planetesimals just a few miles across. The bulk of these icy bodies remain today. We occasionally see them when they journey in toward the Sun and start to shine—as comets.

Extrasolar Planets

It would be statistically impossible for there not to be other planetary systems like our own. The process of planetary formation in the solar system can't be unique, for this would suppose that the birth of the Sun was a freak occurrence, and we know that this was not the case. And there are all those protoplanetary disks that the HST has spotted that are planets in the making. The question is, how do we prove that there are extrasolar planets—planets beyond our solar system? The answer is: with great difficulty, because planets are tiny compared to stars and don't give off light of their own. Nevertheless, astronomers can still detect them.

The first extrasolar planets were discovered in 1991, not around an ordinary star, but around a pulsar—the remnant of a star that blew itself apart. It was not until 1995 that the first planet was found around an ordinary star, 51 Pegasi. Since then more than 200 extrasolar planets have been discovered. They seem to be roughly the same order of size as Jupiter, but usually orbit much closer to their parent star.

Astronomers have not spotted these planets in telescopes, but have detected them indirectly. The main technique they use is to look for stars with a characteristic "wobble." A wobble, or slight motion to and fro, tends to indicate that one or more planets is orbiting the star and affecting it gravitationally. Astronomers detect wobbling stars by examining the spectrum of their light. Their motion causes the dark lines in the spectrum of their light to shift, first toward the blue, then toward the red.

If planets are as common as we think, what are the chances of there being planets like Earth? Good. And could such planets harbor life, intelligent or otherwise? Probably. It is with this in mind that some astronomers are involved in SETI, the search for extraterrestrial intelligence. The giant radio telescope at Arecibo in Puerto Rico is at the forefront of current SETI research (see page 152).

below
Dusty disk
A wide disk of ice and dust surrounds the billion-year old star HD 53143. It could be the equivalent of the solar system's Kuiper Belt of icy rock objects that orbit the Sun beyond Neptune.

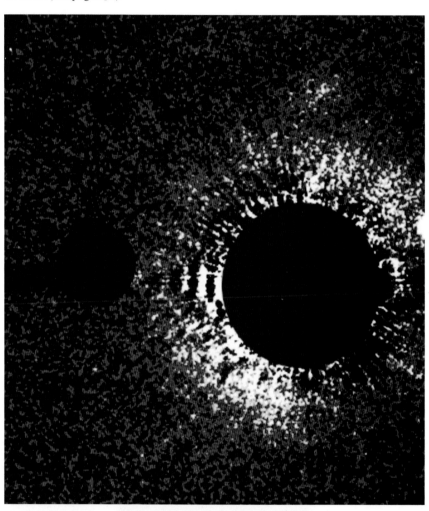

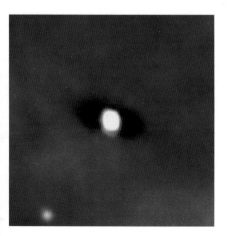
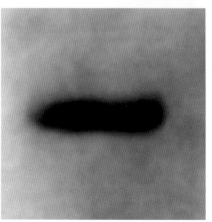
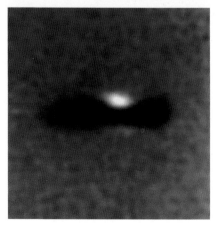

left
Interstellar frisbees
Looking rather like frisbees, these are some of the protoplanetary disks that have been found in the Orion Nebula. They range in size from about two to 17 times the size of our own solar system.

Family of the Sun

It was the priest-astrologers of the early civilizations in the Middle East who laid the foundations of astronomy. They became familiar with the heavens, but had absolutely no idea of what the universe was really like. They did, however, all agree that the Earth was its center.

The belief that the Earth was flat had, by Aristotle's time (in fourth-century BC), given way to the idea of a round or spherical Earth. Aristotle pointed out that the Earth threw a curved shadow on the Moon. Furthermore, the sphere was the perfect shape and therefore appropriate for a body that was the center of the universe. The dissenting voice of Aristarchus of Samos, a next-generation philosopher who argued that the Earth might orbit the Sun, fell on deaf ears.

Much later, around AD 150, the Alexandrian astronomer Ptolemy elaborated the classical Earth-centered view of the universe, which became known as the Ptolemaic system. According to this system, all the heavenly bodies—Sun, Moon, planets—circled around a central Earth. The stars were fixed to the inside of an all-enveloping dark orb, the celestial sphere, which rotated around the Earth once a day.

This concept held sway throughout classical times and the Dark Ages that followed, when astronomy, like so many other

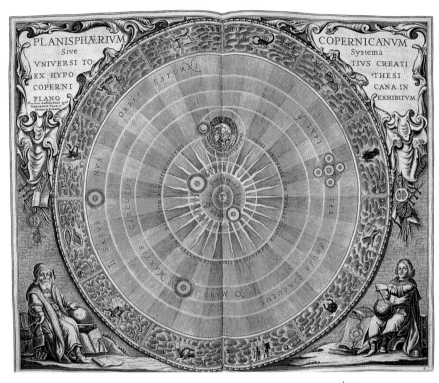

below
The solar system
Orbits of the eight planets and the dwarf planet, Pluto, in our solar system. The planets are widely scattered, traveling within a region of space 8 billion miles (14 billion km) across. The real edge of the solar system, far beyond Pluto's orbit, is marked by an enormous sphere of cometary bodies called the Oort Cloud.

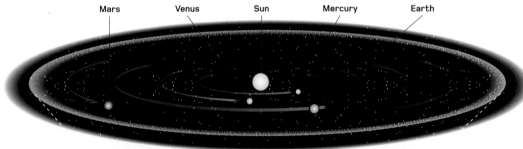

Mars · Venus · Sun · Mercury · Earth

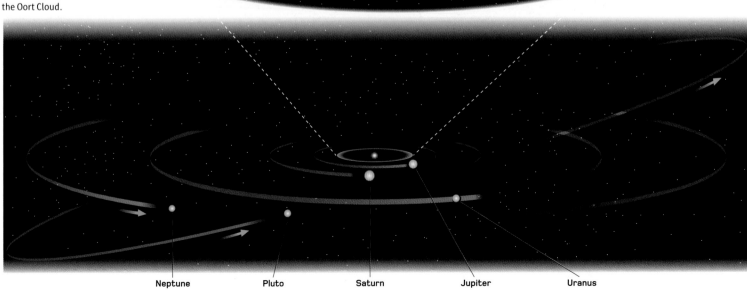

Neptune · Pluto · Saturn · Jupiter · Uranus

branches of learning, suffered almost terminal neglect. Only in the Arab world did the art and craft of astronomy continue to advance. This explains why many stars have Arab names, such as Betelgeuse (in Orion), Algol (in Perseus), and the delicious Zubenelgenubi and Zubeneschamali (Southern Claw and Northern Claw) in Libra. The names of the last two reflect that they were once considered part of Scorpius (the Scorpion).

The Copernican Revolution

In the fourteenth and fifteenth centuries came the rebirth of learning we call the Renaissance. People began to question age-old beliefs, and science (including astronomy) began to advance once more. In 1543, a Polish religious cleric with a passion for astronomy upset the old order and antagonized the Church by putting forward the concept of a Sun-centered, or solar system. His name was Nicolaus Copernicus.

By carefully examining his own and others' observations, Copernicus became convinced that the often bizarre motion of the planets—that they sometimes back-track through the heavens—could better be explained if they and the Earth itself circled the Sun. It took nearly a century for other astronomers, and the Church, to accept the Copernican solar system, which in effect marked the dawn of modern astronomy.

Our Solar System

What is this solar system of ours like? The main bodies are the eight planets and their satellites, or moons. But there are also many other smaller bodies, including dwarf planets, asteroids, Kuiper Belt objects, and comets.

The planets circle round the Sun at different distances. They don't exactly "circle" the Sun, but rather travel around it in elliptical (oval-shaped) orbits. All eight planets orbit around the Sun in much the same plane (flat sheet) in space. They also all travel in the same direction. If you could view the solar system from a point in space above the North Pole, the planets would orbit the Sun counterclockwise. The planets also have another motion: They spin round in space. Earth spins round once in just 24 hours, but the other planets have different spin times, from under 10 hours (Jupiter) to 243 days (Venus).

There are the four terrestrial planets made of rock (Mercury, Venus, Earth, Mars) and four giant gassy planets (Jupiter, Saturn, Uranus, Neptune). The three dwarf planets are Ceres, the largest asteroid; Eris, the largest Kuiper Belt object; and Pluto, which from its discovery in 1930 until 2006 was classified as a planet.

What keeps the planets circling round the Sun? It is the Sun's enormous gravity—enormous because the Sun has such an enormous mass—750 times more than the mass of all the other bodies in the solar system combined. The Sun thus has much better credentials than the Earth for being the center of our corner of the universe.

top
Dusty Mars
The HST captured Mars on 28 October 2005 a day before its closest approach to Earth.

above
Serene Saturn
HST images use four filters to render the colors as seen by the eye through a telescope.

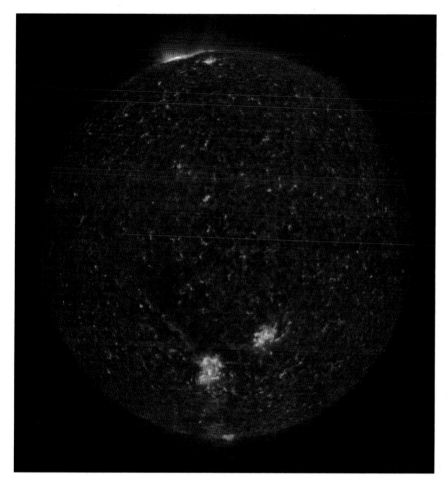

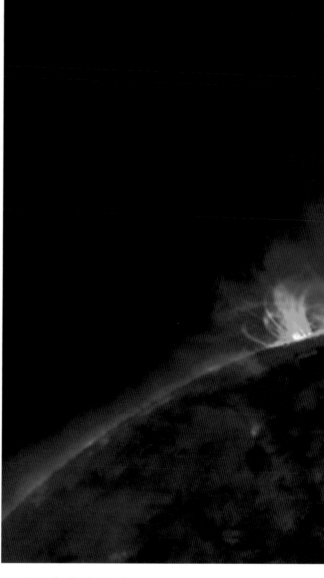

Neighborhood Star

The star at the center of our own solar system, the Sun, is special to us. But in the universe as a whole, it is very ordinary. It is one of at least 500 billion stars in our Galaxy alone. It may seem big to us, but as stars go, it is a dwarf. The biggest supergiant stars such as Betelgeuse in Orion are hundreds of times bigger in diameter. However, for astronomers, the Sun is of extraordinary importance because it is the only star they can study from close quarters. The Sun is a mere 93 million miles (150,000,000 km) away, while the other stars are light-years from us.

Like other stars, the Sun is made up mainly of hydrogen and helium, the two most common elements in the universe. There are also traces of as many as 70 other elements. All these elements are present as plasma, a form of matter in which the atoms exist as ions (charged particles) in a sea of electrons. Matter takes this form at the kinds of temperatures that exist in the Sun and the other stars.

Solar Energy

The temperature of the bright surface of the Sun, the photosphere, is relatively low, about 10,000 degrees Fahrenheit (5,500°C). As you go deeper into the Sun's interior, temperatures rocket. In the core (center), they reach a high of 30 million degrees Fahrenheit (15,000,000°C).

At such astronomical temperatures and equally astronomical pressures, nuclear reactions take place, which produce the energy that keeps the Sun shining. It has been shining steadily for nearly five billion years, and should continue to shine for at least as long again.

In these nuclear reactions the nuclei (centers) of atoms of hydrogen fuse (join) to form nuclei of helium atoms. In the process small amounts of mass seem to be destroyed, but in fact, they are transformed into energy. Albert Einstein quantified this mass-energy transformation in that most famous of mathematical equations: $E=mc^2$, where E is the energy released when a mass m is transformed, and c is the velocity of light. Since c is a huge value — 186,000 miles (300,000 km) per second — the energy released when even a small amount of mass is converted is incredibly large. In practice, four million tons of hydrogen are converted into energy every second.

right
The Sun's crown
In a total solar eclipse, the Moon moves in front of the glaring surface of the Sun and blots out its light. It is then that we can see the Sun's pearly white outer atmosphere, the corona (crown).

far right
Lunar transit
When the STEREO spacecraft pointed its cameras at the Sun on 25th February 2007 they got an extraordinary view. The black disk of the Moon was seen transiting the Sun.

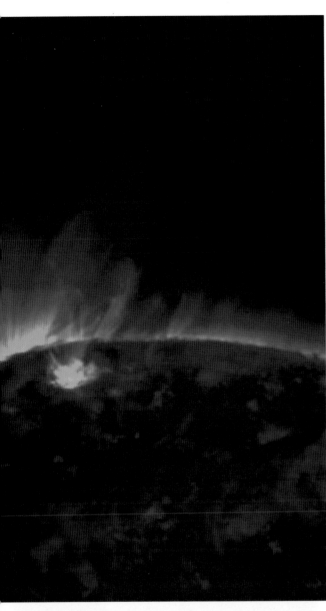

far left
STEREO's Sun
Pictured by the STEREO spacecraft, the Sun is a seething, boiling mass of incandescent gas (mainly hydrogen and helium). The surface looks speckled because of rising (hotter) and falling (cooler) pockets of hot gas.

left
Surface loops
One of the first images sent back to Earth by the STEREO spacecraft, which studies the Sun. This false-color ultra violet view is a close-up of loops in a magnetic active region.

The Sun's Rays

The energy produced in the Sun's nuclear furnace travels to the surface first as radiation, then on convection currents of rising gas. It pours out from the photosphere as visible light and other invisible radiation, such as gamma rays, X-rays, ultraviolet rays, infrared (heat) waves, microwaves and radio waves. These are different kinds of electromagnetic waves—minute electrical and magnetic disturbances in space. They differ in their wavelengths, with gamma rays being the shortest and radio waves the longest.

The HST was designed to study the light and ultraviolet and near-infrared radiation from the distant stars, but not from the Sun. The intensity of the Sun's radiation would effectively burn out all the sensors. Wisely, HST designers equipped the telescope with a protective sunshade.

Essential Sun	
Diameter at equator:	865,000 miles (1,392,000 km)
Average distance from Earth:	93 million miles (149,600,000 km)
Spins on axis in:	25 days at equator
Mass:	333,000 times Earth's mass
Surface gravity:	28 times Earth's gravity
Surface temperature:	10,000°F (5,500°C)
Core temperature:	30 million°F (15,000,000°C)

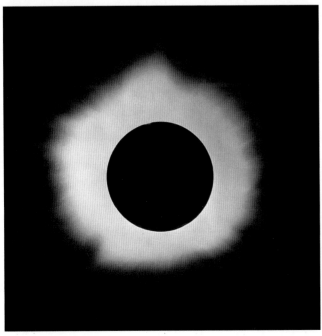

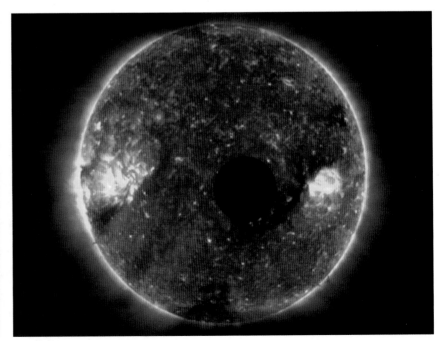

left
The far side
The Apollo 13 astronauts snapped this picture of the Moon's far side on their abortive Moon-landing mission in 1970. The dark region is the Sea of Moscow, the only significant sea on the far side.

The Silvery Moon

One of the oddities of the heavens is that, from Earth, the Sun and the Moon appear to be almost identical in size. Nothing could be further from the truth. With a diameter of 2,160 miles (3,476 km), the Moon is actually only 1/400th the size of the Sun. But, because the Moon is 400 times nearer to us than the Sun, the two bodies appear to be the same size.

Of course, the Moon is quite a different body from the Sun, being made of solid rock rather than incandescent gas. And whereas the Sun shines in its own right, the Moon shines only because it reflects sunlight, though not particularly well.

The Moon is Earth's only natural satellite, orbiting our planet about every four weeks. During this time, because of the geometry of the Sun, Earth and Moon in space, we see more or less of the Moon lit up by the Sun. This makes the Moon appear to change shape in the sky—from a thin crescent, to full circle, and back again. These changing shapes, or phases, of the Moon mark one of the great rhythms of nature.

The Lunar Surface

With our eyes, we can make out two general features on the Moon—dark and light areas. The dark areas were once thought to be seas and were named maria (Latin for "seas," singular "mare"). But in reality they are dusty plains. The lighter areas are highland regions, thought to be remnants of the Moon's original crust.

Many of the Moon's mare regions are circular, and this shape suggests their origin. Early in its history, the Moon was pounded by asteroids, and the largest ones gouged out huge basins from the surface. Subsequently, volcanic activity filled the basins with lava, resulting in the dark, flat landscapes that typify the maria.

The fact that the maria were formed more recently than the Moon's highlands is evidenced by the number of craters. On the maria, craters are few and far between, whereas in the highlands they crowd one another. The largest of these craters, such as Bailly and Clavius, measure over 150 miles (250 km) across.

right
Moon rock
A fine-grained volcanic rock brought back from the Moon by the Apollo astronauts. Similar to terrestrial rocks called basalt, it is riddled with holes where gas bubbled out of cooling lava.

right
Taurus-Littrow
Harrison Schmitt examines a huge split boulder near the Apollo 17 landing site at Taurus-Littrow, on the edge of the Sea of Serenity. As ever, the lunar landscape is hauntingly beautiful.

Thanks to close-up investigation by probes and the exciting expeditions by the Apollo astronauts, we know exactly what the Moon is made of. The surface is covered with a kind of soil that crumbles easily, known as regolith. All the rocks are volcanic in origin. The main types are basalt, much like Earth basalts, and breccia, made up of chips of preexisting rocks cemented together.

Some of the moons that orbit other planets resemble our Moon superficially. Many are heavily cratered from heavy bombardment from outer space. But they have a quite different composition. Most of these moons are made up of varying mixtures of ice and rock—Jupiter's Io is one of the exceptions. It is a yellow-orange volcanic world.

Our Moon is so close that the HST would have to take 130 separate images to cover the Moon's entire face. The telescope is more often used to target moons of distant planets. In May 2005 it discovered two new moons around Pluto.

Essential Moon	
Diameter at equator:	2,160 miles (3,476 km)
Average distance from Earth:	239,000 miles (384,000 km)
Spins on axis in:	27 1/3 days
Circles Earth in:	27 1/3 days
Goes through its phases in:	29 1/2 days
Mass:	1/81 Earth's mass
Surface gravity:	1/6 Earth's gravity

Broom Stars

Of all the objects that light up the night sky, none are as spectacular or as intriguing as comets. Today, when thousands of astronomers are routinely scanning the skies, there is little chance of a comet sneaking up on us undetected, as happened in the past.

In past times, comets seemed to appear suddenly out of nowhere, with their glowing heads and long tails fanning out behind them. To superstitious peoples, the appearance of a comet was a bad omen, a portent of evil, a harbinger of drought or disease, death and destruction. Over 2,500 years ago the Chinese seemed to take a particular interest in comets, which they called broom stars. Ancient Chinese scribes wrote that comets were vile. King Harold of England would probably have agreed, for he was shot in the eye when a comet appeared at the Battle of Hastings in 1066.

A Comet Called Halley

The comet of 1066 was actually making one of its regular appearances in Earth's skies. English Astronomer Royal Edmond Halley was first to recognize this some 600 years later, and so the comet was named after him. Halley's Comet was last seen in Earth's skies in 1986 and will return in about 2061. Thousands of the world's astronomers and a flotilla of spacecraft studied the 1986 appearance of the comet in detail, though the HST had yet to be launched at that time.

Visually, however, the 1986 visit of Halley's Comet was a disappointment. It was faint and difficult to spot among the stars, and observers needed a telescope to see its tail. Comet-watchers had to wait a decade before seeing the first of a pair of spectacular comets with the naked eye. The first was Hyakutake in 1996, the second Hale-Bopp a year later. Hale-Bopp in particular was outstanding—it was one of the great comets of the twentieth century. It hung in northern skies for months, outshining all but the brightest stars. In early 2007, Comet McNaught was also easy to see with the naked eye.

below
Hyakutake
Telescope and HST (inset) images of the 1996 comet Hyakutake. Easily visible to the naked eye, it grew a particularly long tail. Hubble images of the comet focused on the region around the nucleus and showed bits that had broken off.

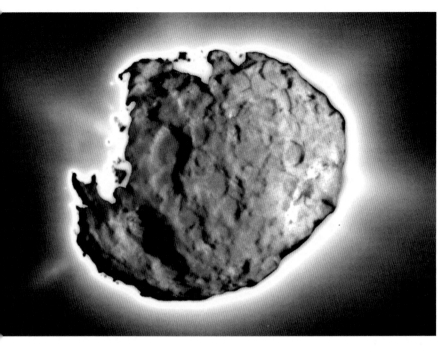

Dirty Snowballs

Astronomers study comets closely, not just because they are so spectacular, but also because they are perhaps the most primitive bodies in the solar system. They seem to originate in a vast reservoir of comets at the outer fringes of the solar system, called the Oort Cloud. This reservoir was created when the planets were being born, and the bodies it contains have remained virtually unchanged since then. They therefore hold the key to understanding the origin and evolution of the solar system.

What are these primitive bodies like? They are made up mainly of a mixture of ice and dust and are often described as dirty snowballs. Far away, in the dark, cold depths of the outer solar system, they remain in deep-freeze. They become visible as comets only when they travel toward the Sun and begin to absorb some of the Sun's warmth.

As they warm up, the icy surface starts to evaporate, releasing clouds of vapor or gas. The gas spurts out in jets, carrying dust

with it. The gas and dust form a cloud around the solid part, or nucleus, of the comet. The cloud reflects sunlight, and the comet becomes visible. The cloud can measure up to hundreds of thousands of miles across, but the nucleus itself is usually only a few miles wide.

The pressure of radiation from the Sun forces dust away from the glowing head of the comet, or coma, and shapes it into a fan-shaped, yellowish tail. Magnetic effects associated with the solar wind make the gas particles glow and force them into a second tail called the ion tail (ions are electronically charged particles). In the biggest comets, the tails can stretch for up to 100 million miles (160,000,000 km).

The HST was used to study comets shortly after it was launched and produced spectacular images of Shoemaker-Levy 9 (see overleaf). It also set its sights on Hyakutake and Hale-Bopp. In April 2006 the HST provided astronomers with extraordinary images of the comet Schwassman-Wachmann 3. The comet was disintegrating before their eyes.

Comet Shoemaker-Levy 9

"I think I've found a squashed comet!" exclaimed US astronomer Carolyn Shoemaker on March 24, 1993. She was examining a photographic plate from a recent exposure during a standard search for comets she regularly conducted at Palomar Observatory in California, with her husband Gene (Eugene) Shoemaker and their colleague David Levy.

In turn, the other two checked the film, they saw a strange apparition with a bar-shaped nucleus, surrounded by a series of little tails and two "wings" of light. None of them had ever seen anything like it, and they were the experts. Together they had already discovered eight comets.

Levy called his friend Jim Scotti at Kitt Peak Observatory in Arizona, who directed his Spacewatch telescope to the position of Carolyn's "squashed comet." Scotti's telescope revealed that it was actually a string of individual comets traveling close together, each with little tails. They were almost certainly pieces of a larger body that had broken up.

The comet-watch trio reported their find, which became known officially as Periodic Comet Shoemaker-Levy 9 (SL9).

In the weeks that followed, astronomers around the world trained their telescopes on SL9. Tracing its path through the heavens revealed that the comet was in orbit around Jupiter. Calculations showed that it must have been the giant planet's powerful gravity that had ripped the larger body apart around July 8, 1993, when it had passed within 13,000 miles (21,000 km) of Jupiter's cloud tops.

Soon, images from a bevy of telescopes, including the HST, showed 21 cometary fragments stretched out in formation — a celestial string of pearls. They seemed to be up to about two miles (3 km) across.

On July 16, 1993, SL9 reached apojove, the most distant point in its orbit around Jupiter, and began to home in on the planet. But calculations showed that this time SL9 would not skim over Jupiter's cloud tops. It would surrender completely to Jovian

above
String of pearls
In May 1994, two months before impact, the 21 icy fragments of Shoemaker-Levy 9 (SL9) are strung out across more than 700,000 miles (1,100,000 km). The HST recorded this image in red light.

above inset
Fireballs
The successive impacts of the comet fragments created huge fireballs in Jupiter's atmosphere, which telescopes on Earth could picture in infrared light.

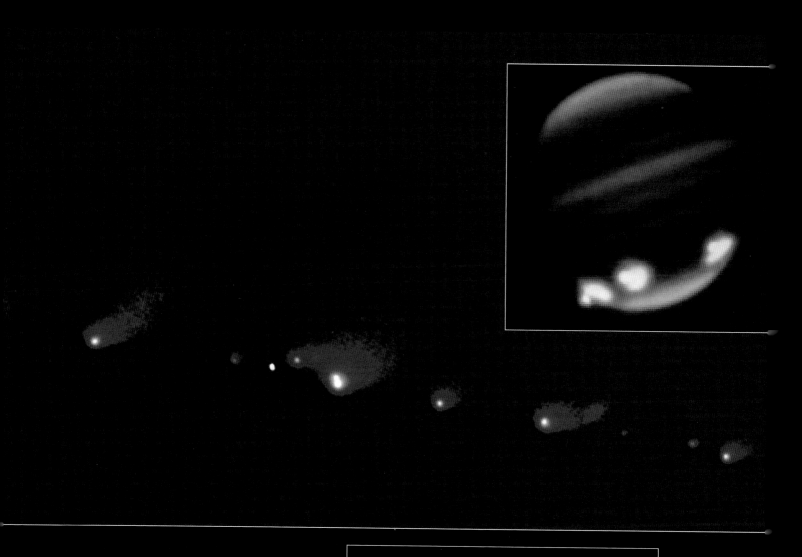

gravity and impact the planet. Astronomers estimated that the first fragment would hit the surface of Jupiter in July 1994.

Sure enough, on July 16, 1994, fragment A plunged into the Jovian atmosphere, creating a fireball that rose 600 miles (1,000 km) above the cloud tops. One by one, over the next seven days, the fragments bombarded the giant planet, creating fireballs and staining the atmosphere with dark clouds.

Most spectacular was the impact of fragment G on July 18, 1994, which sent a fireball leaping 2,000 miles (3,000 km) beyond the cloud tops. It is estimated to have released the energy equivalent to six million megatons of TNT, which made it three billion times more powerful than the atomic bomb that destroyed the Japanese city of Hiroshima in World War II.

This was the first time in history that astronomers had witnessed such a cosmic collision, and they were ecstatic. But it brought home to humanity just how vulnerable planets, including Earth, must be to bombardment from outer space.

left
G impact
This mosaic of images shows the development of the impact site of SL9's G fragment. Just visible in the first image (lower left) is a tiny plume, marking the moment of impact. The second image shows the impact site 90 minutes later. The third shows the site three days later, along with the site of the L fragment impact. The fourth shows the same region two days later still.

Shooting Stars and Asteroids

The Earth is under constant bombardment from outer space, as a night of stargazing should demonstrate. From time to time you may see bright streaks in various parts of the sky. It looks as if stars are shooting across the sky or falling to Earth.

However, the shooting or falling stars you see are meteors—bright streaks made by specks of matter burning up in the atmosphere. Interplanetary space is full of specks and larger lumps of rock and metal, particles we call meteoroids.

Some meteoroids have been chipped off of asteroids or even other planets. Others come from comets, which shed material into space on their passage around the Sun. The intense displays of meteors at certain times of the year, known as meteor showers, are associated with the orbits of particular comets.

The meteoroid particles that cause meteors—usually barely larger than sand grains—burn up to ash, which slowly falls and settles on the ground. Occasionally, bigger meteoroid lumps plummet through the atmosphere, to the accompaniment of more spectacular *son et lumière* (sound and light). The "lumière" can be a multicolored fireball, the "son" a muffled sonic boom. Some of the meteoroid survives and falls to Earth, and becomes what we call a meteorite.

Errant Asteroids

If they are big enough, meteorites can blast out mile-wide craters and pose a real threat to life and limb. But asteroids present a much greater threat to humanity.

Asteroids are a group of much bigger lumps of rock and metal, found mainly in a broad band, or belt, between the orbits of Mars and Jupiter. Even the biggest, Ceres, is less than 600 miles (1,000 km) wide, and only about 100 of the 200,000 discovered so far are bigger than 125 miles (200 km) across. Most are irregular in shape, but the largest are near spherical.

The asteroids appear to be the remains of a failed planet, planetesimals that could not unite because of gravitational disturbances, probably principally from nearby Jupiter.

However, not all asteroids orbit within the asteroid belt. Many wander way beyond it, toward Saturn. Others stray in the other direction and wander among the inner planets. Some pursue orbits that take them uncomfortably close to Earth. These are classed as NEOs, or near-Earth objects.

Historically, Earth has almost certainly been hit by asteroids several times. The impact of an asteroid about 10 miles (16 km) wide is thought to have caused a mass extinction of species, including the dinosaurs, 65 million years ago.

below
Blue streak
A curved streak appears on this image of the Milky Way in Centaurus, which looks toward the center of the Galaxy some 25,000 light-years away. It is a trail made by a tiny asteroid just a few light-minutes away.

The chances are that some time, sooner or later, an NEO will target Earth. The hope is that we shall have sufficient warning to try to do something about it. Several organizations around the world now actively look for NEOs. One is the University of Arizona's Spacewatch, with facilities at Kitt Peak Observatory. Discoveries made by such organizations regularly spark off media hype that "the end of the world is nigh."

below
below
Gaspra
Close up, most asteroids are irregular in shape, like this one, Gaspra. It was the first asteroid imaged at close quarters, by the Galileo probe in 1991. About 12 miles (19 km) long, it is one of the rocky asteroids.

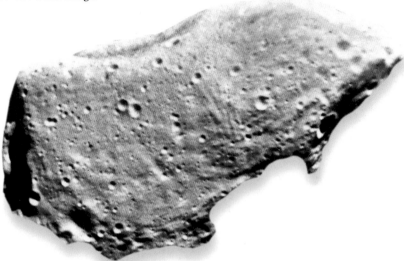

above
Interlopers
Many other asteroid trails have been detected in Hubble images. Most of the bodies are too small to be detected from Earth, and their orbits are unknown.

Close calls

Convincing evidence that Armageddon by asteroids is not just science fiction came within the span of a few weeks in the summer of 2002. First, a NEO (2002 MN), 300 feet (91 m) in diameter, was spotted close to Earth, but receding into space. That was the good news. The bad news was that, three days before, it had missed Earth by less than 80,000 miles (120,000 km).

Not long after, another asteroid (2002 NT7) was spotted much farther away, but it appeared to be making a beeline for Earth, to impact on February 1, 2019. However, subsequent refinement of its orbit suggested that it posed no immediate threat.

Then, a few weeks later, NEO 2002 NY40 flashed past Earth with a close approach of about 300,000 miles (500,000 km). Seen with powerful binoculars, it provided visible evidence that the threat from space is chillingly real for us on Earth.

below
Varied Vesta
The asteroid Vesta has been studied closely by the HST. Vesta is the third largest asteroid, measuring about 340 miles (550 km) across. The color map (bottom) shows the different composition and varied nature of its surface.

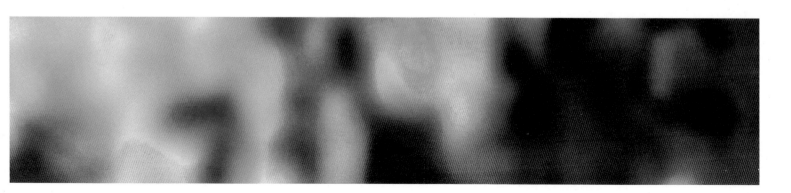

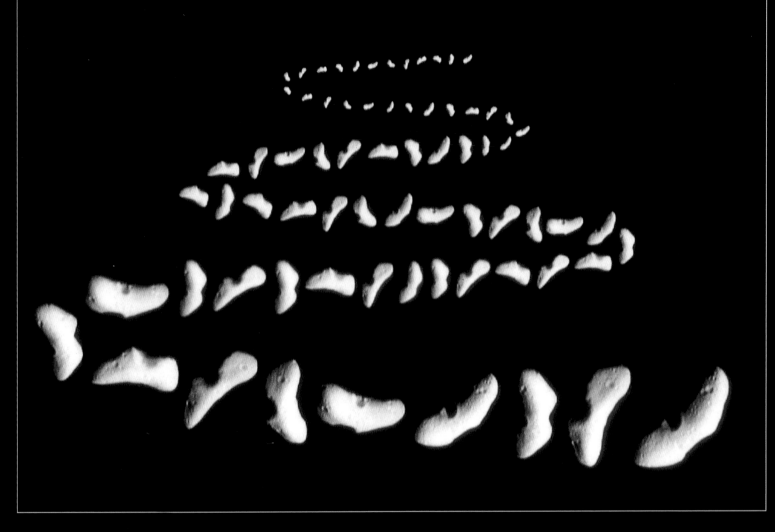

PROFILING:

Eros

Dutch astronomer Carl Gustav Witt discovered asteroid number 433 in 1898. It was called Eros for the son of the Greek love goddess Aphrodite. (In Roman mythology, he was called Cupid and was the son of Venus.)

Eros was the first asteroid found to travel mainly inside the orbit of Mars. It is one of the largest NEOs but fortunately doesn't come much closer to Earth than about 14 million miles (22 million km). Historically, astronomers used observations of Eros early last century to calculate the first really accurate value for the astronomical unit — the distance between Earth and the Sun.

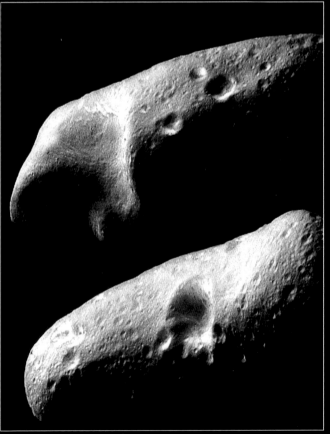

above
The approach
A montage of images taken over three weeks in January and February 2000 as NEAR-Shoemaker homed in on asteroid 433–Eros.

left
Two hemispheres
Two opposite hemispheres of Eros are shown in this pair of images, taken from about 220 miles (350 km). At the left of the top image is the saddle-like feature Himeros. At the center of the lower image is the large crater Psyche.

Even from Earth, astronomers can figure out that Eros is an elongated body that rotates every five hours or so — they can tell this from the way its brightness varies. And they find that it circles round the Sun every 642 days. Eros is one of the most common kinds of asteroids, an S-type; the S standing for silicates, the main mineral ingredients of such asteroids.

We now have a wealth of other data about Eros, for it has been the focus of one of the most incredible feats of interplanetary navigation ever. A spacecraft has not only flown past it, but has gone into orbit around it and then landed. This spacecraft was NEAR (Near Earth Asteroid Rendezvous Mission).

NEAR was launched in February 1996 and took four years to reach its target. Its trajectory took it several times around the Sun and also included a gravity boost from Earth in January 1998. After a failed attempt at orbiting Eros in December 1998, NEAR achieved success on February 14, 2000. It was appropriate that on distant Earth it was the day for lovers — Valentine's Day. Once in orbit around Eros, the spacecraft was renamed NEAR-Shoemaker, for the US planetary geologist and comet-hunter Eugene Shoemaker (see page 124).

NEAR-Shoemaker spent almost exactly a year orbiting Eros at a height of about 30 miles (50 km). It took thousands of close-up images of the asteroid, which proved to be elongated as expected and measured about 21 miles (33 km) long and 8 miles (13 km) across. As well as having a camera for imaging, the spacecraft carried spectrometers to measure the asteroid's composition and a laser range-finder to map its topography.

Like other asteroids that have been imaged, Eros is pockmarked with craters, though it also has smooth regions as well. One of the largest craters, Psyche, is more than 3 miles (5 km) across. Another prominent feature, Himeros, has a saddle shape. Much of Eros is criss-crossed with linear features that were probably created when the tiny world shuddered under impacts with other asteroids. It seems to be covered in a kind of dusty soil, or regolith, presumably produced when the surface rocks are pulverized by meteorite impacts.

During its year-long orbit of Eros, NEAR-Shoemaker greatly exceeded expectations, and its two billion-mile (3,200,000,000 km) mission ended in triumph when it landed on the asteroid's surface on February 12, 2001 — something it hadn't been designed to do. After four-and-a-half hours of de-orbiting and braking manoeuvers, it touched down gently at a sedate four mph (6 km/h).

As the spacecraft was descending, it snapped close-up pictures of the surface from altitudes down to 400 feet (120 m), imaging details as small as half-an-inch (1 cm) across. Mission controllers maintained communications with NEAR-Shoemaker for more than two weeks afterward. Said mission director Robert Farquhar in a masterly understatement: "Things couldn't have worked out better."

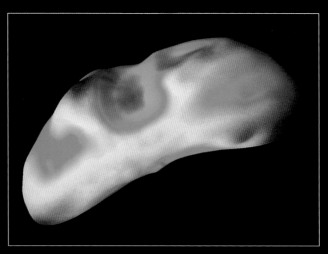

left
Bizarre grades
The elongated shape, density and spin of Eros combine to create curious gravitational topography — uphill and downhill. In this computer model of the asteroid, for example, red areas are uphill and blue areas, downhill.

left
Inside Psyche
The floor of the crater Psyche, pictured from 30 miles (50 km) away. The surface has a thin coating of soil, or regolith, and is peppered with tiny craters.

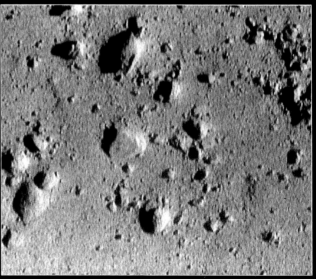

left
Landing mode
NEAR-Shoemaker snapped this image from a range of only 800 feet (250 meters) as it was descending to the surface. The large rocks shown are just a few feet high.

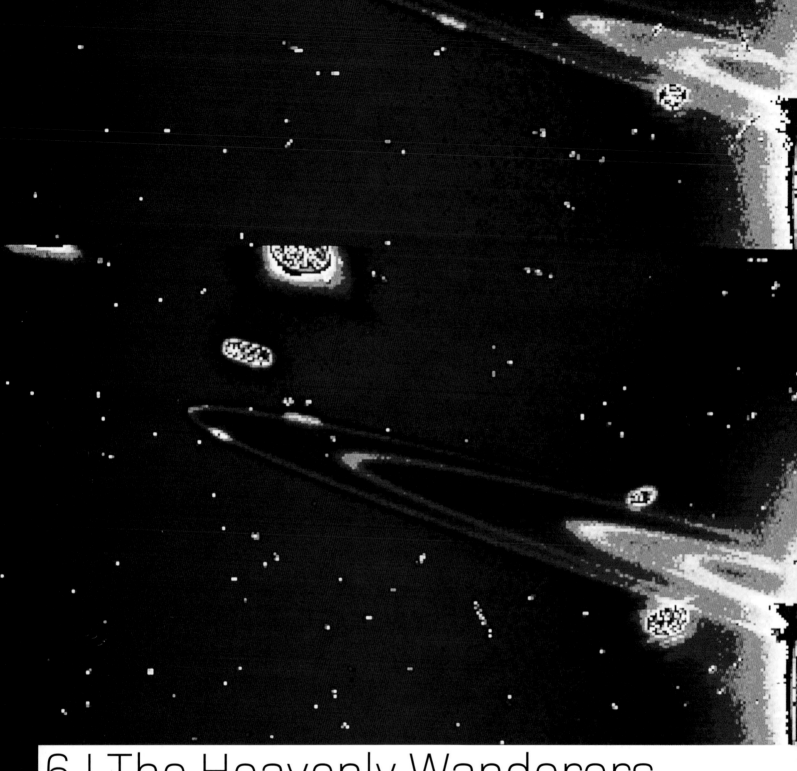

6 | The Heavenly Wanderers
The HST keeps a watchful eye on the planets

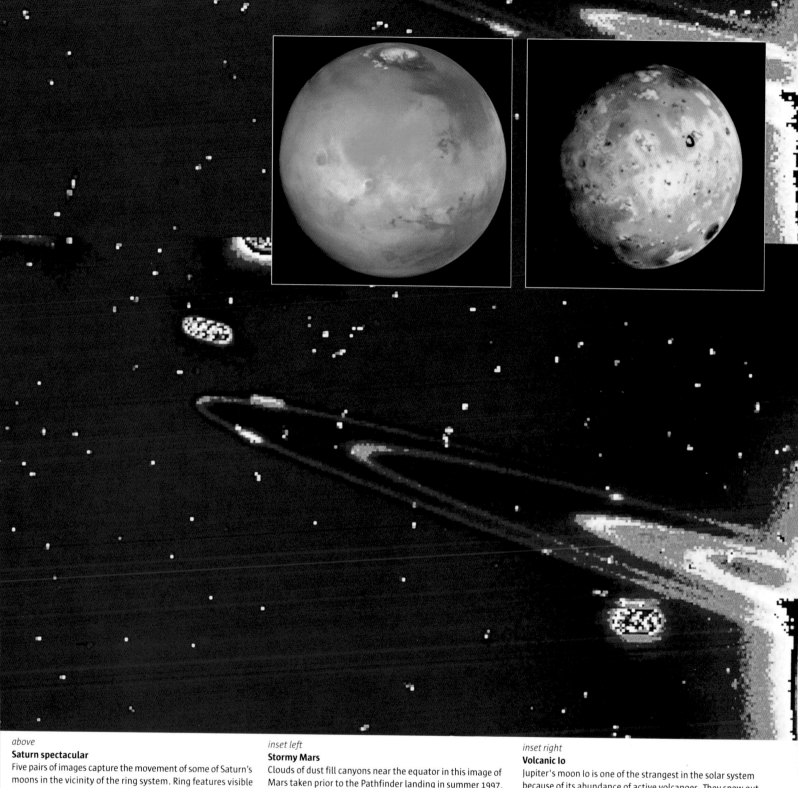

above

Saturn spectacular

Five pairs of images capture the movement of some of Saturn's moons in the vicinity of the ring system. Ring features visible here, from the planet outward, are the C ring, the Cassini Division and the F ring.

inset left

Stormy Mars

Clouds of dust fill canyons near the equator in this image of Mars taken prior to the Pathfinder landing in summer 1997. Clouds of water-ice cover much of the north polar region.

inset right

Volcanic Io

Jupiter's moon Io is one of the strangest in the solar system because of its abundance of active volcanoes. They spew out sulfur to create a colorful landscape that has been likened to the surface of a pizza.

Scorched Worlds

On many evenings of the year, we can see a bright star hanging in the darkening western sky. We call it the evening star. But star it is not—it is the planet Venus.

Of the five planets we can see in the night sky with the naked eye, Venus is the easiest to recognize. Of the others, Jupiter and Mars are the most distinctive, while Mercury is difficult to see because it always remains close to the Sun, and Saturn tends to merge into the stellar background most of the time. The two other planets—Uranus and Neptune—are too faint to be visible at all.

Our knowledge of the planets has been painstakingly gathered over centuries by naked-eye and telescopic observations and more recently by spacecraft, particularly those that have probed solar system space to explore planets and their moons, comets and asteroids at close quarters.

Whereas visits by space probes are fleeting, the HST can keep a constant eye on our planetary neighbors and is always on hand to record transient and unexpected phenomena.

Ironically, observing nearby planets with the HST is a lot trickier than observing the far-distant stars. This is because the planets are moving targets, forever changing their position against the background of "fixed stars" as they orbit the Sun.

As you might expect, the two rocky planets that orbit the Sun at a closer range than Earth—Mercury and Venus—are much warmer. Temperatures on Mercury peak at over 800 degrees Fahrenheit (430°C), and those on Venus can rise as high as 900 degrees Fahrenheit (480°C).

Mercury is so hot because it has a very slow rotation period, taking 59 days to spin around once, compared with Earth's rotation period of just 24 hours. This slow spin, coupled with the planet's 88-day movement around the Sun, means that a point on Mercury's surface stays exposed to 176 days of sunshine at a time. This is what makes temperatures soar. On the other hand, when the Sun sets at a point on Mercury, a night begins that lasts for 176 days. Then, temperatures plummet as low as −290 degrees Fahrenheit (−180°C).

Mercury is too small to have enough gravity to hold onto an atmosphere. This lack of atmosphere not only allows the temperature extremes between day and night, but also makes the planet vulnerable to meteorites. In the distant past, lumps of rocky debris rained down on the planet and gouged out craters large and small. Today the craters remain more or less unchanged, for little erosion takes place on this airless world.

Ground-based telescopes show few details of Mercury's surface. Most of our knowledge about the planet has come from Mariner 10, a probe that made three passes of the planet beginning in March 1974. The HST does not attempt to target Mercury because it is always too close to the Sun.

right
Pockmarked Mercury
This mosaic of images of Mercury returned by Mariner 10 in 1974 shows a planet almost completely covered with craters. Most are billions of years old, dating back to a time of intense meteorite bombardment in the solar system.

left
Lunar vistas
Close up, Mercury's surface resembles some of the more heavily cratered regions of the Moon. A major difference between the two bodies is that Mercury doesn't have any large plains like the lunar seas.

Essential Mercury

Diameter at equator:	3,032 miles (4,880 km)
Average distance from Sun:	36 million miles (58,000,000 km)
Spins on axis in:	58 days 15.5 hours
Circles Sun in:	88 days
Moons:	0

right

Caloris Basin
Circles of mountains ring the biggest feature on Mercury, the Caloris Basin. This was created billions of years ago when an asteroid-sized rock, maybe 60 miles (100 km) across, smashed into the planet.

The Planet from Hell

The planet we know as the evening star, Venus, is rocky like Mercury, but could hardly be more different in most other respects. It is much larger than Mercury, and only a few hundred miles smaller across than our own planet. And like Earth, Venus is surrounded by an atmosphere.

Venus's atmosphere is much thicker than Earth's, with a crushing pressure that is 90 times what we experience. The Venus atmosphere is made up almost entirely of carbon dioxide, the gas that on Earth traps solar energy and causes global warming. On Venus, there is so much carbon dioxide in the atmosphere that it creates a runaway greenhouse effect that has heated up the whole planet. The temperature on Venus is over twice the temperature of your average domestic oven, a temperature that would melt metals such as tin and lead.

The thick clouds that fill the atmosphere permanently hide the planet's surface from our view. However, they are not made up of water droplets like Earth's clouds, but are made up of droplets of sulfuric acid. What a hell of a planet Venus is. If you were to go there, you would be simultaneously oven-baked, crushed and etched to death!

Shaped by Volcanoes

What does Venus look like beneath the clouds? The Venera 9 space probe sent back the first surface picture when it landed

left
Veiled Venus
In ultraviolet light, characteristic cloud patterns show up in Venus's atmosphere. Ordinary light can't penetrate the thick clouds that prevent us from seeing the surface.

right
Lifting the veil
The Magellan probe used radar to penetrate Venus's cloud cover and image the surface. The bright region in this global mosaic of images is the larger of the planet's two continents, Aphrodite Terra.

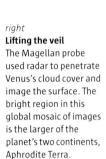

below left
Soaring volcanoes
Maat Mons is one of the most impressive of Venus's many volcanoes. Around 4 miles (6 km) high, it is named for an ancient Egyptian love goddess. Most features on the planet have been given female names. For example, there is a plateau named Guinevere, a chasm named Diana and a crater named Cleopatra.

in 1975, showing a landscape strewn with scattered rocks. Radio telescopes operating in radar mode began to reveal the general topography of the planet at much the same time. Radar can reveal surface features because it uses radio waves, which can penetrate clouds.

It was a radar probe named Magellan that completed the first high-resolution survey of Venus's surface between 1990 and 1994. Orbiting around the planet, it returned amazing images, revealing virtually a whole planet shaped by volcanic activity. Hundreds of volcanoes cover the landscape, many of them miles high, and some of them could be active.

The low-lying plains that make up more than four-fifths of the surface of Venus have been formed by repeated flows of lava spewed out from the volcanoes. Most of the volcanoes are of the shield type found widely on Earth, which have a particularly runny lava that can flow a long way.

There are only two main highland regions on Venus, which we can liken to continents. The largest, Aphrodite Terra, lies near the equator and is about as big as Africa. The other, Ishtar Terra, is about Australia's size and lies to the north.

Venus, like Mercury, always stays relatively close to the Sun. The HST has only made occasional observations of the planet, when Venus has been at its greatest elongation (its furthest point to the east or west of the Sun).

The HST hasn't observed the surface of Venus because it doesn't operate at microwave or radar wavelengths. It has, however, revealed features of the atmosphere by observing at ultraviolet wavelengths.

Essential Venus	
Diameter at equator:	7,521 miles (12,104 km)
Average distance from Sun:	67 million miles (108,000,000 km)
Spins on axis in:	243 days
Circles Sun in:	224.7 days
Moons:	0

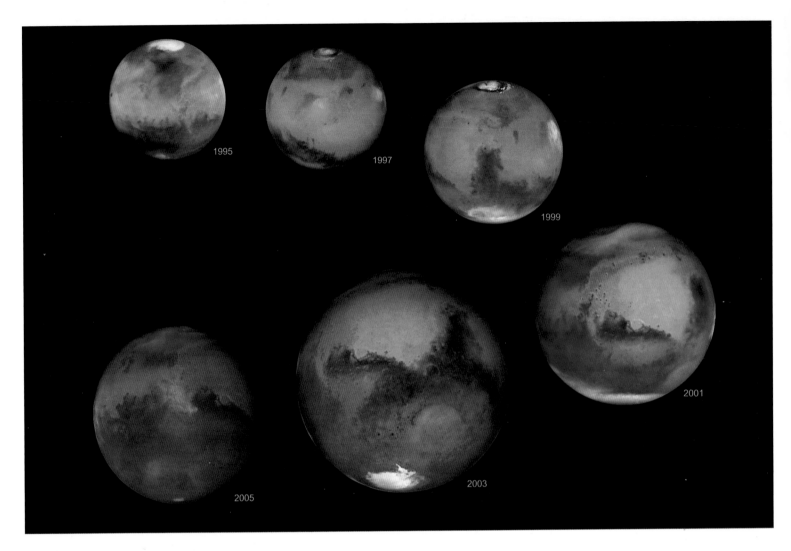

1995

1997

1999

2001

2003

2005

The Red Planet

After Venus, two planets vie for the title of the brightest: Mars and Jupiter—the next two planets beyond Earth, going away from the Sun. They can both shine more brilliantly than any star, but we can easily distinguish them. Whereas Jupiter shines a brilliant white, Mars has a distinct reddish-orange hue, which accounts for its popular name, the Red Planet.

It was the color of Mars that prompted its name in classical times. Likening its color to that of blood and fire, the Romans named it after their god of war.

Through a telescope you can see vague, dark features on Mars, and there are ice caps that come and go. Careful observation reveals that the planet rotates in a little over one Earth day. Space probes have revealed clouds in the Martian atmosphere.

An atmosphere, clouds, a day of similar length to our own, ice at the poles, dark markings (could they be vegetation?). Surely this must be a planet like Earth? Maybe it also has intelligent life? Many people a hundred years ago believed this to be the case.

Prominent among these believers was US astronomer Percival Lowell, who founded the Lowell Observatory in Arizona specifically to study Mars. He was convinced that the "canals" he thought he saw on the surface were being built by a dying race of Martians, trying to channel water from the ice caps to farmland near the warmer equator.

The reality is quite different. A series of space probes from 1965 to the present day—from Mariners and Vikings to Mars Express, Exploration Rovers, and Reconnaissance Orbiter—have

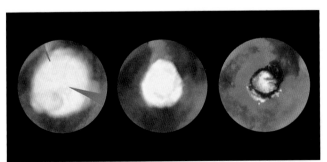

left
In retreat
The extent of the polar ice caps on Mars changes with the seasons (which are nearly twice as long as Earth's). These HST images show how the northern ice cap shrinks over a period of five months, between early spring and early summer.

Water and Weather

There may be no canals on Mars, but there are channels that look much like dried-up riverbeds on Earth. They were almost certainly created by flowing water long, long ago. We know that today the only water on Mars is in the form of ice, in the polar caps and the crystals that make up the clouds. Recent spacecraft like Mars Odyssey have also detected the presence of vast quantities of water ice just below the surface—a kind of Martian permafrost.

If water did once flow on Mars, and astronomers are convinced that it did, then the planet must once have had a much milder climate than it does now—one in which water could exist as a liquid. If so, could life have gained a toehold in that milder climate? Maybe. Perhaps the first human explorers of the Red Planet will find out by unearthing fossils of ancient Martian life. We shall have to wait and see.

The HST has shed no light on the prospects of life on Mars. But it has provided a wealth of information about the Martian atmosphere and weather, closely monitoring the changes over the seasons of the Martian year, which is nearly twice as long as our own.

shown Mars to be a barren planet, covered in vast deserts and pockmarked with craters. There is only a trace of atmosphere, and temperatures are generally much lower than those on Earth; they rise above freezing only near the equator in mid-summer. There are no signs of canals, or vegetation or life of any kind—intelligent or otherwise.

Mars's Grand Canyon

Mars does, however, boast a natural linear feature—a great gash in the planet's crust that runs along the equator. This ancient geological fault system in the planet's crust is named Valles Marineris, or Mariner Valley, after the Mariner spacecraft that discovered it. It begins near the other outstanding natural feature on Mars—the bulge of the Tharsis Ridge, which carries four of the biggest volcanoes we know in the solar system.

Valles Marineris runs for nearly 3,000 miles (5,000 km), in places widening to more than 250 miles (400 km) and plunging as deep as 4 miles (6 km). This vast system of canyons dwarfs Earth's Grand Canyon in Arizona.

Essential Mars	
Diameter at equator:	4,220 miles (6,792 km)
Average distance from Sun:	142 million miles (228,000,000 km)
Spins on axis in:	24 hours, 37 minutes
Circles Sun in:	687 days
Moons:	2 (Phobos and Deimos)

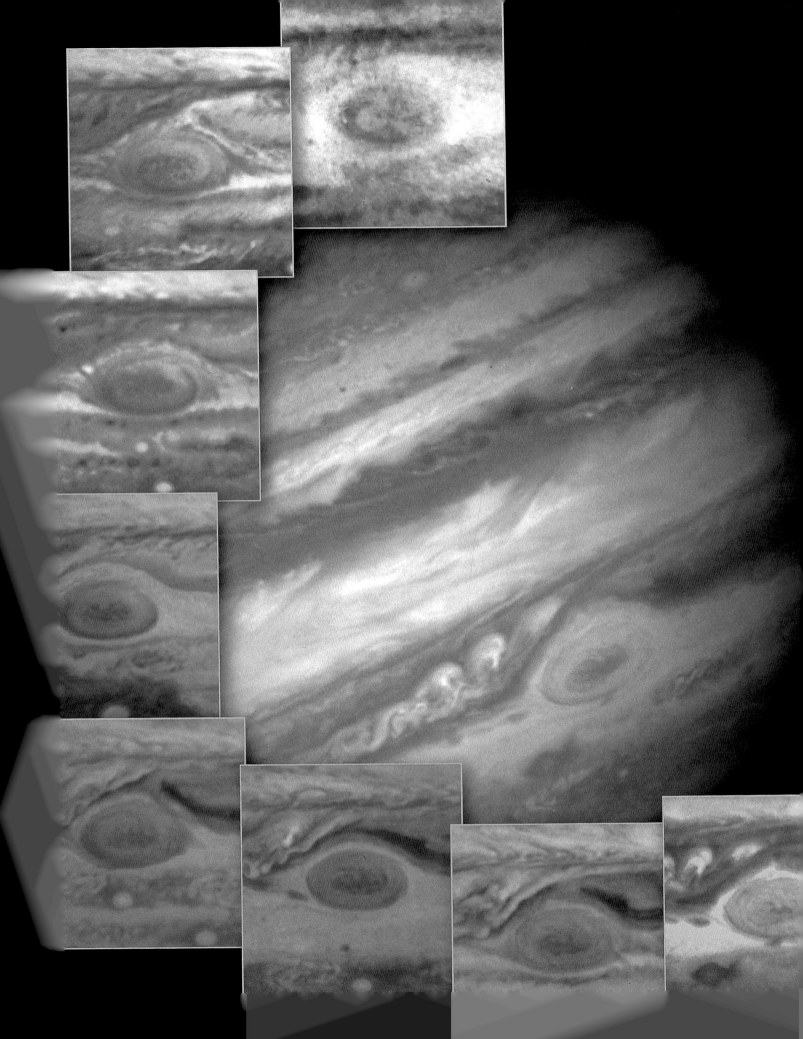

Gigantic Jupiter

left
Changing spots
The HST keeps a close eye on giant Jupiter and images the hectic activity taking place in the colorful cloud bands superbly. It has in particular been monitoring the marked changes in appearance of the planet's super-hurricane, the Great Red Spot.

right
Eclipse on Jupiter
Three black circles on Jupiter's face are shadows cast by three of Jupiter's largest moons. Ganymede is the blue circle at upper right and its shadow is on the planet's left edge. Io's shadow is at left and Callisto's is on the right edge.

With 11 times the diameter of Earth and with more mass than all the other planets put together, Jupiter is truly gigantic. Even though it never gets closer to us than about 400 million miles (600,000,000 km), it shines brilliantly in the night sky. This is due not only to its enormous size, but also to its cloudy atmosphere, which reflects light well.

Jupiter is quite a different body from the terrestrial planets. Its atmosphere is much deeper and is composed mainly of hydrogen, with some helium. There is no solid surface, no rocky landscape, beneath Jupiter's atmosphere.

Pressures at the foot of the atmosphere rise so high that they compress the hydrogen gas into liquid, causing the whole planet to be covered by an endless ocean of liquid hydrogen. More than 12,000 miles (20,000 km) below the ocean surface, pressures soar so high that they compress the very atoms of hydrogen. The hydrogen turns into a kind of liquid metal, rather like the liquid metal mercury we find on Earth.

Saturn has a similar makeup to Jupiter, and both planets are often referred to as gas giants. So are the next two planets out, Uranus and Neptune, though they have a somewhat different composition. All four planets have only a relatively small core of rock at the center.

The Stormy Atmosphere

Jupiter has the most fascinating atmosphere of all the planets. It is criss-crossed with bands of colorful clouds, and many other kinds of features—swirls, eddies, plumes and ovals. Astronomers call the dark bands belts and the light ones zones.

The clouds have been drawn out into these parallel bands by Jupiter's rapid rotation. Strangely, the biggest planet rotates the fastest, spinning round once in space in a little under 10 hours. The various bands don't all travel at the same speed; those in the equatorial region travel fastest. And they don't even travel in the same direction—some travel east, some west. The resulting interaction between them sets up violent eddies and turbulence that creates the wavelike features and ovals that cover Jupiter's disk.

Mostly, these features are constantly changing and transient. But one persists: the famous Great Red Spot (GRS). First glimpsed more than three centuries ago, the GRS appears to be an enormous and permanent super-hurricane.

The HST has studied Jupiter's atmosphere routinely; it is capable of returning high-resolution images every hour and a half. It has studied the planet both in the visible spectrum and at invisible wavelengths. Ultraviolet studies reveal, for example, Jupiter's aurora—light displays in polar regions, similar to the Northern and Southern Lights we see in Earth's polar regions.

The Galilean moons of Jupiter—the big four of at least 63 Jovian satellites—have also come under the HST's close scrutiny. A prime target has been Io, often called the pizza moon because of its colorful appearance. Io is the most dynamic moon in the solar system. Its surface is undergoing constant change due to erupting volcanoes. These volcanoes do not spew out molten rock like volcanoes on Earth, but molten sulfur instead.

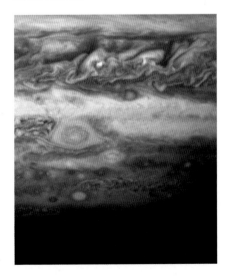

left
Red Spot Jr
The HST has witnessed the birth of Jupiter's second red spot (center left), roughly half the size of the Great Red Spot. Nicknamed Red Spot Jr, it formed as three white oval storms that merged 1998–2000. This view shows it in April 2006.

below
The big four
The HST has taken a "family portrait" of Jupiter's four large Galilean moons. Ganymede is the largest—indeed, it is the largest moon in the solar system, larger even than the planet Mercury. Callisto is only slightly smaller.

Ganymede

Callisto

Io

Europa

Essential Jupiter	
Diameter at equator:	88,850 miles (143,000 km)
Average distance from Sun:	483 million miles (778,000,000 km)
Spins on axis in:	9 hours, 55 minutes
Circles Sun in:	11.9 years
Moons:	63

The Ringed Wonder

Encircled by a set of bright shining rings, the gas giant Saturn remains a firm favorite among astronomers. The other gas giants (Jupiter, Uranus, and Neptune) also have ring systems, but they are but a pale imitation of Saturn's and cannot readily be seen from Earth.

With a similar makeup to that of Jupiter, Saturn has a banded atmosphere, but the parallel cloud bands are much less distinctive. Not so much activity takes place in the atmosphere, though furious storms do rage from time to time.

Ring-a-Ring

Saturn's rings reflect light brilliantly and make the whole planet much brighter than it otherwise would be. In the night sky, it can outshine all the stars except Sirius and Canopus, but for only some of the time. Even when it is farthest away from us—at a distance of nearly 900 million miles (1,400,000,000 km)—it is still as bright as a first-magnitude star.

The amount of extra brilliance provided by the rings varies. Because of the inclination of Saturn's spin axis, on Earth we see the rings at different angles during the almost 30 years it takes Saturn to orbit the Sun. Roughly every 15 years, the rings all but disappear, when they appear edge-on to our line of sight. The rings are edge-on in the years 2009 and 2025.

Rings and Shepherds

In telescopes, we can make out three main rings from Earth: the A, B, and C rings, from outer to inner. Brightest is the B ring, which is separated from the A ring by the apparently

below
Changing aspects
Saturn's incomparable rings present different aspects to us year by year. These HST images were taken between 1996 (bottom), a year after the ring-plane crossing, and 2000 (top), when the rings were nearly fully open.

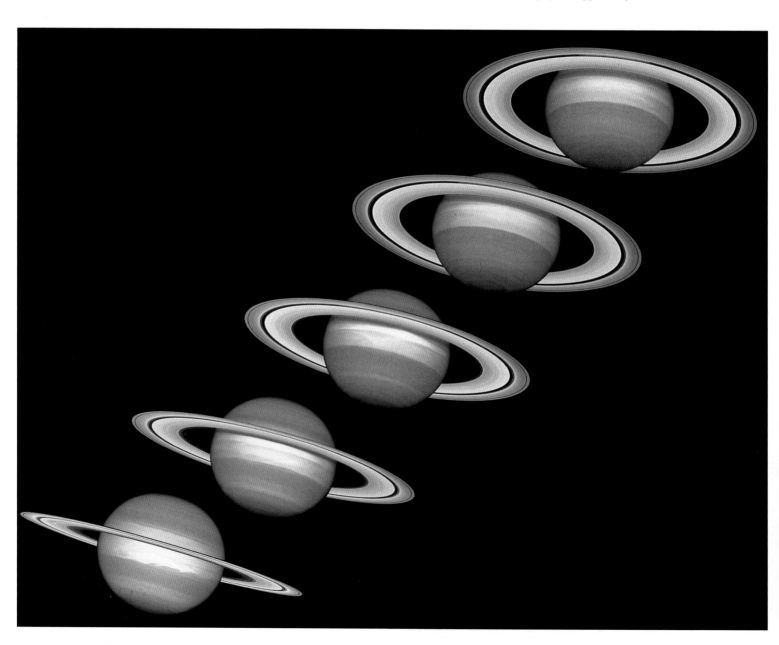

ringless Cassini Division. In total, the ring system measures some 170,000 miles (272,000 km) across.

Spacecraft, particularly the two Voyagers and Cassini, have changed our perspective on the rings. They have spotted more rings inside and outside those visible from Earth. It turns out that each set of rings is made up of thousands of individual ringlets. The ringlets are our perception of particles of material whizzing in orbit around the planet at high speed. These particles are made up mainly of ice, which is why they reflect light so well. Some particles are as small as sand grains; others are as big as boulders.

The camera eyes of the Voyager probes provided an explanation of why the particles remain in the rings. They detected a number of tiny moons among the rings that by their gravity seem to keep the ring particles in place. They are called shepherd moons, because they seem to "herd" the particles like a shepherd herds sheep.

As with Jupiter, HST observers have concentrated on monitoring the changes in Saturn's atmosphere. They have logged the progress of many a storm, after being alerted by astronomers on the ground. This is an example of the growing cooperation between terrestrial and space astronomers.

The HST has also sets its sights on Saturn's planet-sized moon Titan. Bigger than Mercury, Titan is unique among moons because of its thick atmosphere, which astronomers believe might resemble Earth's early atmosphere in its chemical composition.

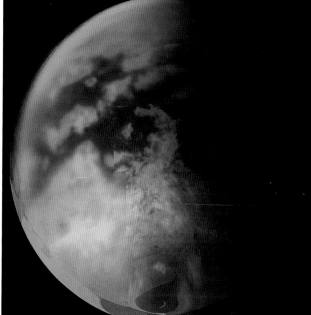

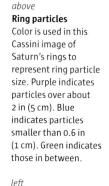

above
Ring particles
Color is used in this Cassini image of Saturn's rings to represent ring particle size. Purple indicates particles over about 2 in (5 cm). Blue indicates particles smaller than 0.6 in (1 cm). Green indicates those in between.

left
Titan's surface
A 50:50 mix of rock and water ice, Titan's surface was shaped by Earth-like processes. The central feature in this Cassini image of December 2006 may be an old impact basin. Mountain ranges are to its south east.

right
Titan's atmosphere
Haze in Titan's atmosphere (blue), which is made up mainly of nitrogen, like Earth's. Surprisingly, the pressure of the atmosphere is 50 percent higher than it is on Earth.

Essential Saturn	
Diameter at equator:	**74,900 miles (120,500 km)**
Average distance from Sun:	**888 million miles (1,429,000,000 km)**
Spins on axis in:	**10 hours, 39 minutes**
Circles Sun in:	**29.5 years**
Moons:	**56**

right
False-color Saturn
It took the HST 20
exposures with the
WF/PC over a period
of 20 hours to produce
this full image of
Saturn. The rings are
lit by the Sun from
underneath, and it is
the least dense
sections of the rings
that let most light
through. The brightest
features we see,
therefore, are the
outer F ring, the Cassini
Division and the inner
C ring.

Infrared Saturn
To celebrate the HST's eighth birthday, the science team "gift wrapped" everybody's favorite planet in vivid colors. It was imaged in the infrared by NICMOS — the first time this instrument had been turned on the planet.

right

Ring-plane crossing
Two images of Saturn taken three months apart. The upper image (August 1995) records the ring-plane crossing — the time when the plane of the ring system crosses our line of sight; the next crossing will take place in 2010.

left

Saturn's lights
Saturn's polar auroras show up brilliantly in ultraviolet light. They are just as spectacular as Jupiter's.

New Worlds

On the night of March 13, 1781, William Herschel was studying the stars in Gemini at his home in Bath, England. German-born, he had neglected his profession as a musician to become an enthusiastic astronomer. Among Gemini's stars he came across an object that intrigued him. "A curious nebulous star or perhaps a comet," he recorded. But this body was no star or comet; it was a new planet, Uranus.

When astronomers determined the planet's orbit, they were astonished. This new wandering star pursued an orbit that took it 1,800 million miles (2,900,000,000 km) from the Sun. This was twice as far away as Saturn, regarded since ancient times as the farthest planet. Herschel's discovery literally doubled the size of the known solar system, and triggered the search for more new worlds.

Small Giant

Uranus is the third-biggest planet after Jupiter and Saturn, but it is less than half the size of Saturn. The strangest thing about the planet is that its spin axis is tipped right over. The axis of spin for all the other planets is more or less at a right angle to the plane of its orbit around the Sun. But Uranus's spin axis is nearly in the orbital plane itself.

Uranus appears as a featureless, bluish-green orb. Like the other gas giants, its deep atmosphere is made up mainly of hydrogen and helium. There are also significant traces of methane, the gas we use to cook with on Earth. It is this gas that is responsible for the color of the atmosphere, because it absorbs red wavelengths from sunlight, producing a blue color.

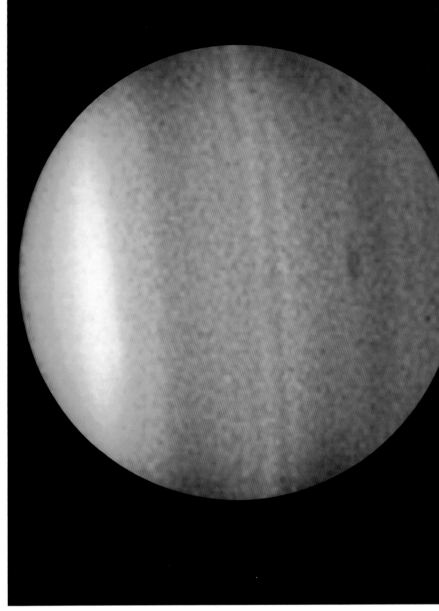

Surprise, Surprise

In 1977, Uranus sprang another surprise. Astronomers setting out to chart the occultation of a star by the planet discovered rings around it. In 1986, Voyager 2 pictured the rings more clearly, showing that there are 11 in all. The HST discovered a second set of two giant rings in December 2005.

Another intriguing Voyager discovery came when the probe closed in on Miranda, the smallest Uranian moon we can see from Earth. Miranda's surface displays geological features that are unique in the solar system. Quite different landscapes abut one another with no gradual transition between them.

above
Dark spot
In 2006, the HST recorded a dark spot about two thirds the size of the USA in Uranus' northern hemisphere. Previous HST images showed no evidence of the spot.

left
Crescent Uranus
A thin crescent of Uranus, pictured by Voyager 2 as it completed its encounter with the topsy-turvy planet.

Essential Uranus	
Diameter at equator:	31,760 miles (51,120 km)
Average distance from Sun:	1,786 million miles (2,875,000,000 km)
Spins on axis in:	17 hours, 14 minutes
Circles Sun in:	84 years
Moons:	29

Patchwork Landscape

It is possible that such geological patchwork might have come about as a result of a catastrophic collision with another body eons ago. The collision would have smashed the moon to pieces, and then gravity would have caused the moon to reform again, with all the pieces of different geology jumbled together haphazardly.

Voyager 2 also spied a number of tiny moons, invisible to us from Earth. Among them were two shepherd moons that orbit on either side of the outer, Epsilon ring. Named Cordelia and Ophelia, they are both only about 20 miles (30 km) across.

The HST has carried out regular observations of Uranus, its rings and its moons. It has spied more activity in the atmosphere than did Voyager 2, following the changing patterns of clouds and haze. When the Advanced Camera for Surveys (ACS) detected the new rings in 2005, it also discovered two small moons, Mab and Cupid.

Blue Planet

The enthusiastic search for other planets triggered by Herschel's discovery of Uranus came to fruition in 1846. On September 23 of that year, German astronomer Johann Galle discovered planet number eight, which was subsequently called Neptune. It proved to be only a fraction smaller than Uranus. Recent observations have shown that the two planets are remarkably similar, even though their orbits are more than a billion miles (1,600,000,000 km) apart.

Neptune has relatively more methane in its hydrogen/helium atmosphere, and as a result the planet appears a deeper blue than Uranus. But both planets seem to have a similar ocean of warm water, liquid ammonia and methane underneath the atmosphere.

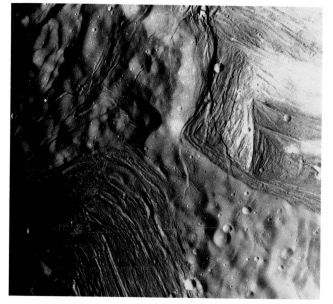

above
Epsilon's ringlets
Many individual ringlets make up Uranus's outermost ring, the Epsilon, pictured here by Voyager 2. The Epsilon is also the broadest ring, measuring about 60 miles (100 km) across.

above
Amazing Miranda
The strange patchwork landscape of Miranda, where different kinds of terrain butt together — cratered hilly regions, chevron or V-shaped areas, and oval grooved regions that look like a gigantic racetrack. There is no geology like it anywhere else in the solar system.

right
New rings
The HST's sharp view has uncovered a pair of faint giant rings girdling Uranus and its previously known ring system. It has also spied two new moons, Mab and Cupid.

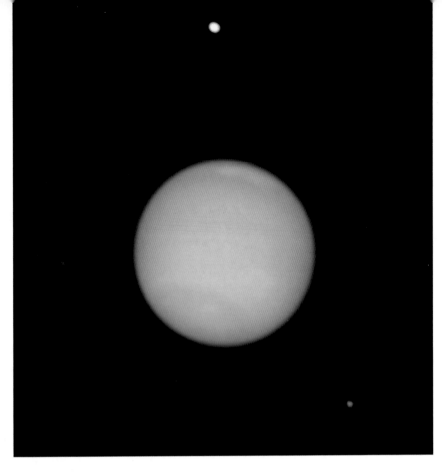

Neptune's Rings and Moons

Like Uranus, Neptune has a ring system, but it is even sparser and fainter. Two of the rings are relatively bright, and the other three very faint indeed. The two brightest are named Adams and Leverrier, after the two mathematicians who independently figured out where the eighth planet could be found (Englishman John Couch Adams and Frenchman Jean Urbain Leverrier). There is also a sixth partial ring.

On its flyby of Neptune, Voyager 2 discovered six new moons, adding to the two that can be seen from Earth, Nereid and Triton. One was Proteus, bigger than Nereid but impossible to see from Earth, because it orbits so close in that it gets lost in the planet's glare. Of the remaining five new moons Voyager discovered, four orbit within the ring system, presumably acting as shepherds for the ring particles.

When the HST first trained its instruments on Neptune, it revealed no trace of the Great Dark Spot Voyager had seen. But it spotted plenty of other cloud activity in the atmosphere.

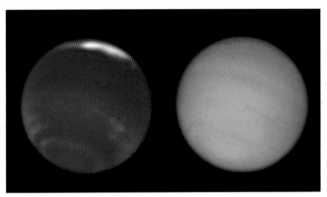

Ice Worlds

With Neptune, we come to the end of the progression of gas giants that dominate the outer reaches of the Solar System. Then, as we penetrate ever deeper into space, we encounter planetary bodies that we call the ice worlds.

We don't even have to leave Neptune to find one. Neptune's largest moon Triton is one. Measuring some 1,700 miles (2,700 km) in diameter, it is made up of rock and ice. Its surface is covered with frozen gases, such as nitrogen and methane, as well as ice. The most curious feature of Triton's surface is its volcanoes. Totally unlike Earth's molten lava volcanoes, those on Triton emit a kind of slushy, frozen gas-ice in a process termed cryovolcanism. Geyser-like eruptions also puff out gas mixed with dark particles, which form dark streaks on the pale ice and snow.

above
Natural world
This natural color view of Neptune was produced by combining HST images in red, green, and blue light. It is one in a series taken in 2005 that show a dynamic atmosphere. Clockwise from the top are four of Neptune's moons; Proteus, Larissa, Despina, and Galatea.

right
Changing weather
The HST monitors the rapidly changing weather on Neptune. Atmospheric features become visible in enhanced-color views (right) and in images taken using methane filters (left).

Neptune's Clouds and Storms

When Voyager 2 flew past Neptune in 1989, it revealed that the blue planet's atmosphere was much more active—had more weather—than Uranus's. Being much farther from the Sun, Neptune receives less solar energy, which is usually what drives planetary weather systems. So you might expect that Neptune's atmosphere would have little weather.

The truth is quite the reverse. Voyager's cameras spotted distinct bands in the atmosphere, rather like Jupiter's and Saturn's. There were also oval storm-type features; a particularly large one became known for obvious reasons as the Great Dark Spot. There were also high-level, wispy clouds, rather like the cirrus clouds we get in Earth's atmosphere. But Neptune's wispy clouds seem to made up of crystals of methane rather than water ice.

Why is Neptune's atmosphere more active than that of Uranus? It seems that the planet has an internal heat source, which drives the planet's weather systems. Internal heating would also explain why at the cloud tops, both Uranus and Neptune have more or less the same temperature, of about −350°F (−210°C).

Essential Neptune	
Diameter at equator:	30,780 miles (49,530 km)
Average distance from Sun:	2,800 million miles (4,505,000,000 km)
Spins on axis in:	16 hours, 6 minutes
Circles Sun in:	165 years
Moons:	13

right
Icy geysers
Neptune's biggest moon Triton is in deep freeze. It has the coldest surface (−390°F/ −235°C) of any body we know in the solar system. The image shows the south polar region, which seems to be covered with pinkish snow. The dark streaks show where fine dust is erupting from icy geysers.

Pluto

Astronomers still continued to try to find new planets long after Neptune had been discovered. Prominent among them was Percival Lowell, noted for his belief in intelligent Martian life. At the Lowell Observatory in Arizona, Clyde Tombaugh eventually found a planetary body in February 1930. It was named Pluto.

Pluto was considered the ninth planet of the solar system from its discovery until 2006. It was then the smallest and most distant planet. But during the closing decades of the twentieth century, its planetary status was increasingly questioned by some astronomers.

This small rock-and-ice body is unlike its large, gassy planetary neighbors. It follows a very elongated orbit and it is inclined to the orbital plane of the planets. And for 20 years of its 248.6-year orbit Pluto is closer to the Sun than Neptune. This last occurred between 1979 and 1999. Pluto also has a moon, Charon, that is about half its own size.

Pluto has not been visited by spacecraft and is comparatively little known. New Horizons is on its way and will arrive in 2015. Pluto is much too remote for ground-based telescopes to spy surface features. But spectroscopic analysis has shown that it is covered with frozen nitrogen, methane, and other gases. The HST has been invaluable in the study of Pluto. It can separate Pluto and Charon, and in May 2005 it discovered two additional small moons, Nix and Hydra.

above
Pluto revealed
Pluto is a tough target even for the HST. But this is one of the best ever images of Pluto and its moons. The large one to the right is Charon. The two smaller ones are Nix (top) and Hydra.

above right
Eris
The HST provided the first images of distant tiny Eris that could be used to determine its size. It found Eris is only a little larger than Pluto.

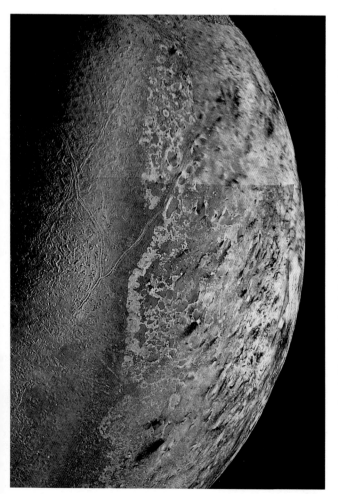

Dwarf Planets

In October 2003, an object larger than Pluto was found, and astronomers felt that Pluto could no longer be regarded as a planet. In August 2006, a new class of object — called dwarf planets — was introduced by the International Astronomical Union, the astronomer's professional world body, and Pluto was to be the prototype. Dwarf planets, like planets, orbit the Sun, and have sufficient mass and gravity to be nearly spherical, but, unlike a planet, a dwarf planet has not cleared the neighborhood around its orbit.

There are three known dwarf planets and more are expected. In addition to Pluto, there are Eris and Ceres. Eris is the largest; its size is not certain but it is accepted to be between a few tens of miles and a few hundred miles larger than Pluto, which is 1,432 miles (2,304 km) across. Eris and Pluto both belong to the Kuiper Belt, a flattened belt of comet-like rock-and-ice bodies encircling the solar system beyond the orbit of Neptune. The first of these were discovered in the 1990s; over 1000 are now known and it is expected there are vast numbers more.

Ceres is the largest asteroid and orbits the Sun within the Main Belt between Mars and Jupiter. It is made mainly of rock with water ice. The Dawn spaceprobe is scheduled to arrive at Ceres in 2015 and reveal this world to us.

above
Ceres
Visible and ultraviolet observations made by the HST are combined to produce this color image of Ceres. Brighter and darker regions could be asteroid impact features.

Background Briefing

Introducing Telescopes | The Hubble Space Telescope

above
VLT telescopes
The four 27-foot (8.2-m) reflectors that make up the Very Large Telescope (VLT) at the Parana Observatory at Atacama in Chile. The telescopes are named for sky objects in the native Chilean dialect. In the foreground is Yepun; behind it (from the left) are Antu, Kueyen and Melipal.

Introducing Telescopes

We can explore the universe with just our eyes and view the constellations, meteors, comets, and eclipses. However, as an optical instrument, the eye has a great disadvantage. It has a small aperture (opening) — the pupil — to let the light through. The principal instrument that astronomers use, the telescope, has a much greater capacity for light-gathering.

The Italian genius Galileo heard of the newly invented telescope in 1609; he quickly constructed his own and trained it on the heavens. He became the first to see the mountains on the Moon and the phases of Venus. The four large moons of Jupiter that he saw are called the Galilean moons in his honor. Galileo used a set of lenses to gather and focus light from the heavens. Many amateurs still use this kind of telescope, called a refractor. An objective lens at the front gathers and focuses incoming light, and the observer views the image formed through an eyepiece.

However, most astronomers use a telescope with mirrors, called a reflector. A curved, parabolic mirror (the primary) is used to gather and focus the light into an image, which is then viewed through the eyepiece. The instrument favored by amateur astronomers is called a Newtonian reflector, because it uses the same configuration as the reflector built by Isaac Newton around 1672. A secondary mirror reflects light gathered by the primary mirror into the eyepiece at the side. In an alternative configuration called the Cassegrain, a secondary mirror reflects light into the eyepiece through a hole in the primary. This is the configuration used by the HST.

The progress of astronomy since Galileo's time is a result of the improvement in the size and design of telescopes. The bigger the telescope, the more efficient it is in gathering light, and the more clearly it can resolve (distinguish) objects.

William Herschel built the finest instruments of his day and discovered a new planet, Uranus, in 1781. Lord Rosse built a gigantic instrument he called the Leviathan and was first to spot the spiral nature of M51 (the Whirlpool Galaxy) in 1845. With the groundbreaking 100-inch (2.5-m) Hooker telescope, Edwin Hubble proved in the 1920s that there were star systems — galaxies — beyond our own and that the universe appeared to be expanding.

Very Large Telescopes

Today's foremost instruments use very large mirrors. The two Keck telescopes at the Mauna Kea Observatory in Hawaii have mirrors 33 feet (10 m) across. The four instruments that make up the Very Large Telescope (VLT) in Chile each have mirrors 27 feet (8.2 m) across.

The mirrors in these telescopes are not made of a single piece of glass. It would be difficult to control and adjust a mirror of that size without creating distortions. These mammoth mirrors are made in segments, which merge together to make one big mirror. Each segment is individually supported and controlled by a computer that constantly adjusts it so that, all together, the segments always form the perfect shape. This technique is known as active optics.

Individually, each of the four telescopes of the VLT is superb, and when they work together as they are designed to, they produce spectacular images that match the quality of those from the HST. When they work in unison, along with three smaller 6-foot (1.8-m) instruments, they create an effective telescope 390 feet (120 m) across.

left
Keck domes
The domes of the twin Keck telescopes on the summit of Mauna Kea, Hawaii. Inside are the world's largest optical and infrared telescopes. Each is eight storeys tall and has a mirror more than 30 ft (10 m) in diameter.

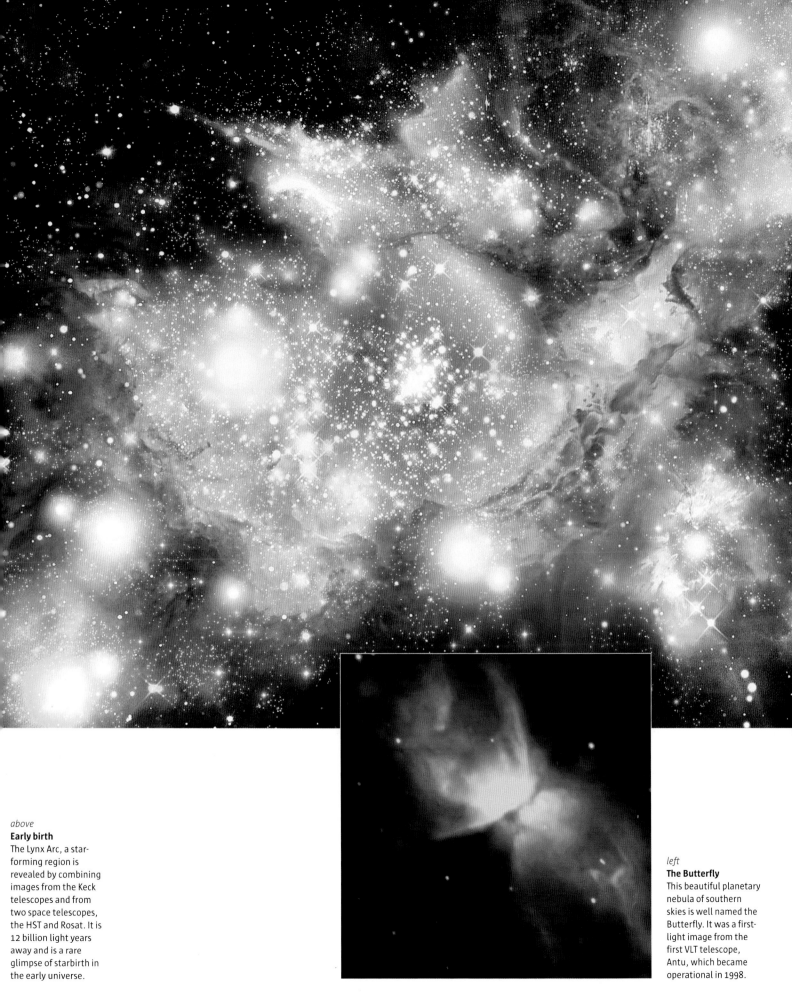

above
Early birth
The Lynx Arc, a star-forming region is revealed by combining images from the Keck telescopes and from two space telescopes, the HST and Rosat. It is 12 billion light years away and is a rare glimpse of starbirth in the early universe.

left
The Butterfly
This beautiful planetary nebula of southern skies is well named the Butterfly. It was a first-light image from the first VLT telescope, Antu, which became operational in 1998.

The Invisible Universe

In 1931, a communications engineer named Karl Jansky from Bell Laboratories in New Jersey, was trying to identify interference he was getting on his radio equipment. After ruling out any local sources, he realized what was happening. The interference was coming from the heavens. That observation proved to be the springboard to one of the most exciting branches of astronomy—radio astronomy. It led astronomers to discover whole new kinds of heavenly bodies, including quasars, pulsars and radio galaxies.

Astronomers tune into the radio waves that emanate from the heavens with radio telescopes that look nothing like their optical counterparts. A radio telescope usually takes the form of a huge metal dish. The dish collects incoming radio waves and focuses them on a central antenna. The signals then pass to a receiver. After amplification, they are fed to a computer for analysis and display as a false-color image, with different colors being assigned to different radio intensities and wavelengths.

Big and Versatile

The biggest single radio telescope is located near Arecibo in Puerto Rico. Measuring 1,000 feet (300 m) across, it is built into a natural bowl in a hilltop. Since it is fixed, it must rely on the Earth's rotation to direct it toward different parts of the sky. A much more versatile instrument is the Very Large Array near Socorro, New Mexico. It is made up of 27 separate dishes, each 82 feet (25 m) across. The dishes are mounted on rail tracks and can be moved into various configurations. Together they present a collecting area some 20 miles (30 km) across.

Radio telescopes at different locations can also be linked to simulate even larger dishes. This is called VLBI (very long baseline interferometry). Theoretically, it can create an effective collecting area with the diameter of the Earth.

Invisible Astronomy

The radio waves that pour down on Earth from the heavens come from what is often called the invisible universe—the universe we can't see because our eyes can't detect it. As witnessed by radio images, the invisible universe can look nothing like the visible one. Still, there is more to the invisible universe than just radio waves.

We detect objects in the heavens by the energy they emit. Visible light and invisible radio waves are just two forms of energy we receive from heavenly bodies such as stars and galaxies. They also give off energy as invisible gamma rays, X-rays, microwaves, and ultraviolet and infrared rays. All these rays are kinds of electromagnetic radiation, which differ only in wavelength.

above
Parkes RT
One of the largest radio telescopes in the Southen Hemisphere, at the Australian National Radio Observatory, near Parkes, New South Wales. Its dish measures 210 feet (64 m) in diameter.

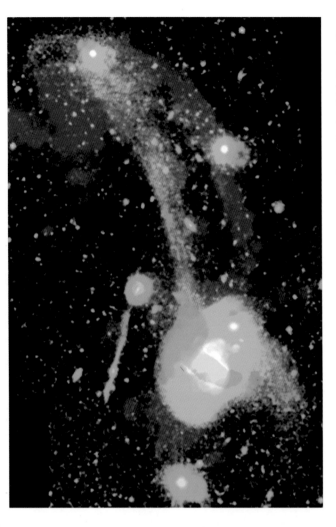

In order to get a comprehensive picture of the heavenly bodies, astronomers ideally need to study them at all of the wavelengths they give out. On the ground, astronomers can study light and radio waves from the heavens because they pass through the atmosphere. But they can't study other wavelengths because the atmosphere absorbs them. This is one of the reasons that astronomers launch telescopes and other instruments into space.

Another reason for launching telescopes into space is that, orbiting hundreds of miles high, they are above Earth's polluted atmosphere. The atmosphere, in effect, offers us a dirty window on space. It is full of dust, clouds, water vapor and shimmering air currents, all of which distort the faint light that reaches us from the heavens.

Space Observatories

Spacecraft have been returning information about the space environment from the very beginning of the Space Age. Launched on January 31, 1958, the first US satellite, Explorer 1, discovered intense donut-shaped bands of radiation around the Earth, which are called the van Allen belts. The first dedicated astronomy satellite, OAO-2 (Orbiting Astronomical Observatory-2) went into orbit 10 years later.

Since that time, scientists have launched dozens of astronomy satellites—space observatories—that span the full range of invisible wavelengths (except for radio waves). From shortest to longest, they cover gamma rays, X-rays, ultraviolet rays, infrared rays and microwaves. The HST, while primarily used for visible light, also covers some ultraviolet and near-infrared wavelengths.

left
Radio view
The Very Large Array reveals the peculiar galaxy Arp 299. It is the result of two spiral galaxies colliding and merging. A long tail (in blue) extends from the main body of the two.

below left
Cassiopeia A
Data from the HST is combined with observations from two more space telescopes, Spitzer and Chandra to produce a stunning image of Cassiopeia A. It is the remnant of a supernova explosion.

below right
Spitzer
An artist's rendition of the Spitzer Space Telescope. In orbit above Earth, it observes the universe in the infra red.

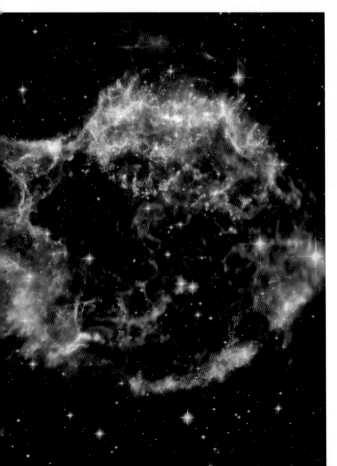

Gamma Rays

Gamma rays have the shortest wavelengths and pack the most energy. They are given off by the most violent activities that take place in the universe. Scientists have detected them coming from pulsars and quasars and in incredibly energetic bursts from unknown events in distant galaxies. Other gamma-ray sources have been identified as events in which ordinary matter and its mirror-image opposite, antimatter, annihilate each other.

The US Compton Observatory, launched in 1991, provided the first comprehensive view of the gamma-ray universe. Europe's Integral (International Gamma-Ray Astrophysics Laboratory), is designed to provide unprecedented high-resolution imaging of gamma-ray sources. It operates in a looping orbit that takes it as far as 95,000 miles (153,000 km) from Earth. In this orbit it spends most of the time outside the potentially damaging van Allen radiation belts in a relatively undisturbed space environment that is ideal for long-duration, real-time observations. Integral was launched in 2002 on a two-year mission, now extended to December 2010.

X-Rays

The wavelengths of X-rays are longer than those of gamma rays and shorter than those of ultraviolet rays. They still pack a lot of energy, which is why they can penetrate body tissue. In the heavens, X-rays are emitted by very hot objects—with temperatures of millions of degrees. Typical X-ray sources include the solar corona (the outer atmosphere of the Sun), supernova remnants and the gas that spirals around black holes. Key X-ray observatories include Rosat (Roentgen satellite, launched in 1990), Chandra (1999) and XMM-Newton (1999).

Ultraviolet Rays

Ultraviolet wavelengths are just shorter than those of the violet light of the visible spectrum. Some ultraviolet (UV) gets through Earth's atmosphere — it is the radiation that tans us. Thankfully, most UV is absorbed by a layer of ozone in the upper atmosphere. The hottest stars are best observed in the ultraviolet because they give out most energy at these wavelengths. The International Ultraviolet Explorer (launched in 1978) provided outstanding data for 18 years. GALEX (Galaxy Evolution Explorer), launched in 2003, is investigating how star formation evolved from the early universe up to the present. It is also identifying objects for further study.

left
Black hole
An x-ray image taken by Chandra of galaxy NGC 4696 shows a vast cloud of hot (red) gas surrounding a bright area that harbors a supermassive black hole.

above
Einstein's Andromeda
X-ray sources in the Andromeda Galaxy, imaged by the Einstein Observatory (launched 1978), originally designated HEAO-2. Einstein was the first large X-ray telescope, which pinpointed more than 5,000 sources.

right
Whirlpool
X-ray and ultraviolet images taken by XMM-Newton are combined in this view of the Whirlpool Galaxy (M51). Red indicates low energy, blue, high. Compare it with the HST view on page 89.

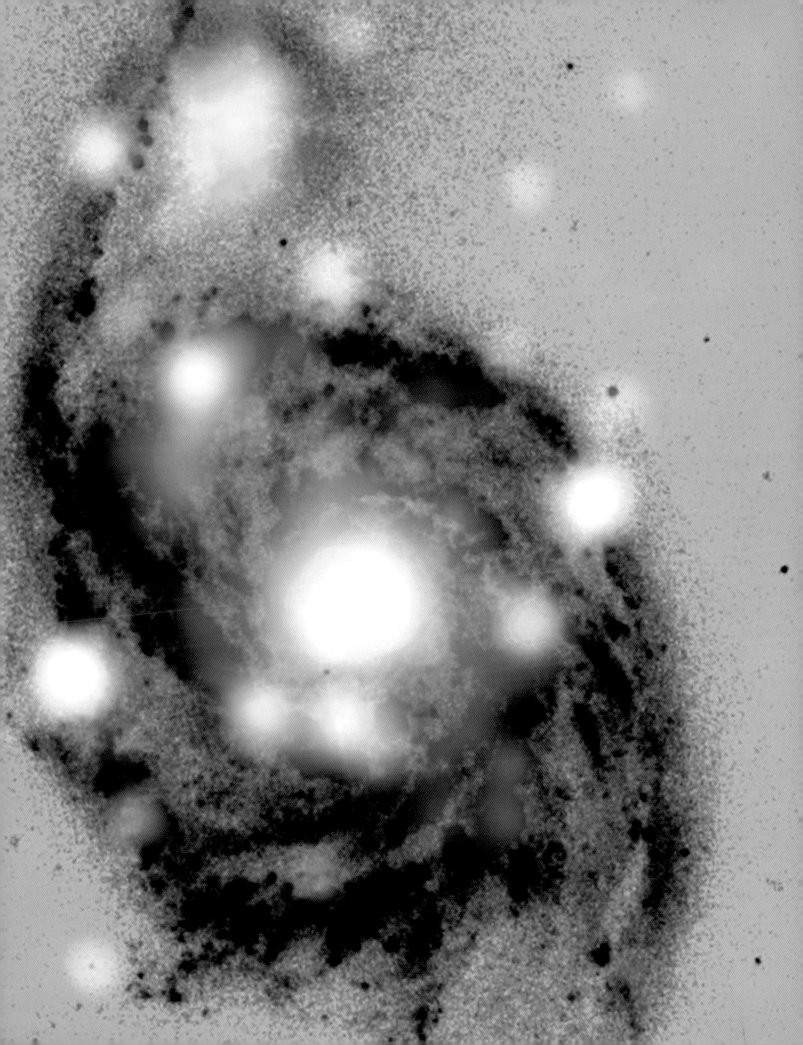

Infrared Rays

Infrared rays have slightly longer wavelengths than visible red light. We receive heat from the Sun in the form of infrared rays. Some infrared wavelengths reach the ground and can be studied from mountaintop observatories. But most infrared wavelengths are completely absorbed by the atmosphere. Airborne observatories such as SOFIA (Stratospheric Observatory for Infrared Astronomy) study the infrared wavelengths from high altitudes. But the most spectacular work has been done by IRAS (Infrared Astronomy Satellite, launched in 1983), ISO (Infrared Space Observatory, launched in 1995) and Spitzer, which, since its launch in August 2003, has been studying a wide variety of astronomical objects, from the solar system to the distant reaches of the early Universe.

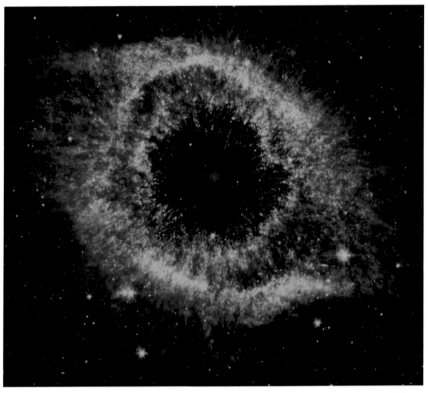

Microwaves

Microwaves have the shortest radio wavelengths. They are used terrestrially to relay communications signals and in cooking with microwave ovens. COBE (Cosmic Background Explorer), launched in 1989, used microwaves to survey the cosmic background radiation of the whole sky. It detected slight "ripples," or variations, in the radiation, regarded as the thermal "echo" of the Big Bang. COBE's findings supported current thinking on the Big Bang and cosmological theory.

Solar Observatories

By studying the Sun, we find out what ordinary stars are like — for there are billions of stars like it in our Galaxy alone. And there are also more pressing reasons for studying this, our local star, because what happens in the Sun can affect Earth and change our "space weather."

As with the other stars, we study the Sun both from the ground and from space. On the ground, astronomers use quite different instruments. Solar telescopes produce images by projection, using mirrors. The McMath solar telescope at Kitt Peak Observatory in Arizona is the world's biggest, with a distinctive triangular design.

The astronauts on the Skylab space station made the first comprehensive study of the Sun from space in 1973–74. Using a series of eight instruments, they probed the Sun at many different wavelengths and achieved spectacular results. More recently, Solar Max (short for Solar Maximum Mission) was launched in 1980 to monitor the Sun at a time of solar maximum, when sunspots and other solar activities reach their climax.

Ulysses (launched in 1990) embarked on a circuitous route via Jupiter to spy on the polar regions of the Sun, which had never before been investigated. SOHO (Solar and Heliospheric Observatory) began continuous observations of our star in 1995. The satellite Hinode (formerly Solar-B) is equipped with three solar telescopes and has been observing the Sun from soon after its launch on 22 September 2006.

above
Helix Nebula
In Spitzer's infrared view of the Helix Nebula the red center denotes the final layers of gas blown off by the central dying star. Infrared light from the outer layers of gas is represented in blues and greens.

right
Leftovers
A whole sky map produced from data returned by COBE, the Cosmic Background Explorer (launched 1989). The map shows tiny temperature variations in the radiation left over from the Big Bang.

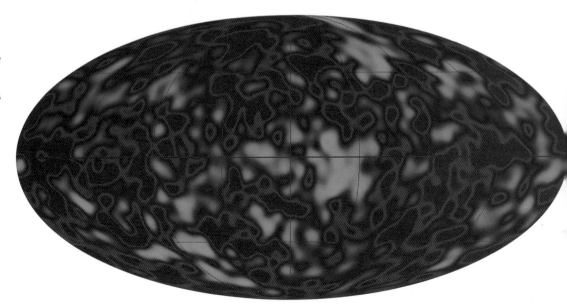

right
Solar reflections
The 11-story-high McMath solar telescope at Kitt Peak National Observatory in Arizona. Sunlight captured by a mirror (heliostat) on top is reflected down the inclined tunnel deep underground, then back up to an observation room, where an image is formed.

left
Hinode's Sun
Hot gas leaps high above the Sun in this view taken by the Solar Optical Telescope aboard the Hinode spacecraft, in November 2006. The gas rises from a sunspot on the Sun's surface.

Probing the Planets

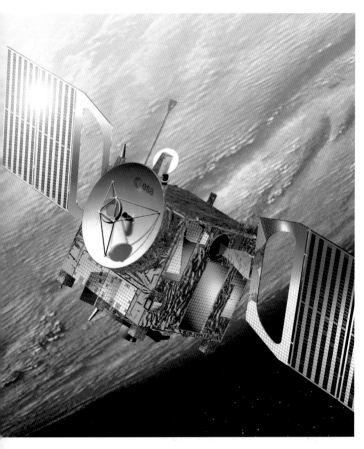

For most of the Space Age, scientists have launched astronomy satellites to study the distant stars. To study the planets and other members of the solar system, they launch different kinds of spacecraft — space probes. Unlike satellites, probes are designed to escape completely from Earth's powerful gravity. To do so, they must be launched at what is called the escape velocity, which on Earth is a speed in excess of some 25,000 miles per hour (40,000 km/h).

Although it missed its target planet, Venus, by thousands of miles, the probe Mariner 2 made the first deep-space discovery in 1962. It revealed that the planet had a heavy carbon dioxide atmosphere. Three years later, Mariner 4 returned the first close-up pictures of a planet, showing the cratered landscape of Mars.

As the years passed, the cameras of these tiny robot explorers revealed the mysteries of one planet after another — Venus, Jupiter, Saturn, Uranus, and Neptune. To date, the dwarf planet Pluto has not been visited by any space probes, though one is now on its way. In January 2006, New Horizons started its journey to Pluto. On arrival in 2015, it will start five months of study before traveling on into the Kuiper Belt.

The planets are not the only targets of space probes. Giotto and other spacecraft carefully encountered Halley's Comet in 1986. Galileo returned the first images of an asteroid (Gaspra) in 1991. NEAR-Shoemaker landed on the asteroid Eros in 2000, in an amazing feat of interplanetary navigation (see page 128).

left
Venus Express
An artist's expression of the European space probe Venus Express in orbit around Venus. It arrived in April 2006 and is expected to work until 2009.

below
Opportunity
The rover Opportunity landed in a highland plain, Meridiani Planum. Pale dirt was spattered over the Martian surface as it landed; at left and center of the image is its discarded heatshield.

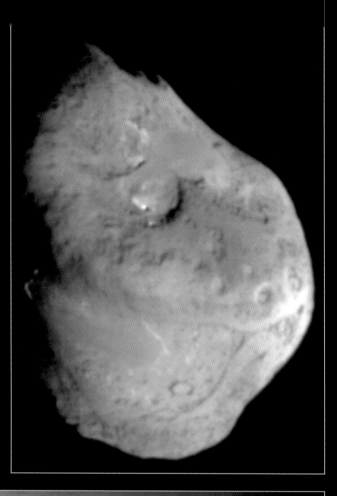

right

Cometary encounter
The nucleus of Comet Tempel I is seen by the Deep Impact spacecraft as it approaches. Five minutes after this image was recorded, Deep Impact's impacting probe smashed into the comet's surface.

far right

Europa
The Galileo space probe made 11 close flybys of Jupiter's moon Europa. Images returned show an icy surface crossed by dark lines, and cracks and ridges thousands of miles long.

below

Jupiter and Io
Io, Jupiter's moon, is seen above the planet's colorful surface. The image was taken by Cassini on the dawn of the new millennium, 1st January 2001, as it flew on to encounter Saturn.

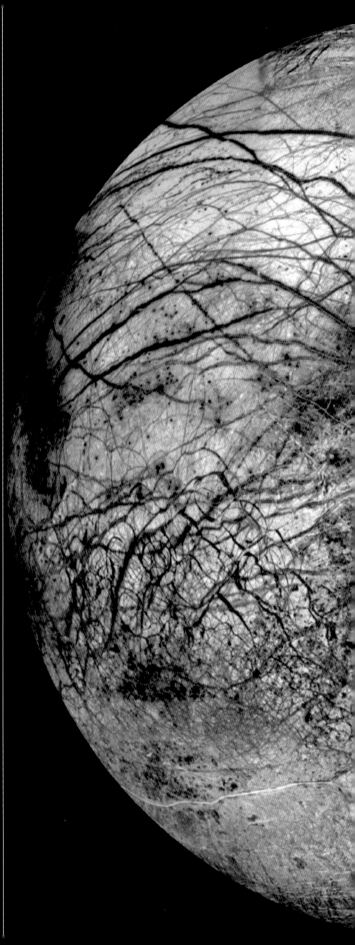

left
Lift-off!
An Atlas V rocket lifts off from its launch pad at Kennedy Space Center in Florida on January 19 2006. Onboard is New Horizons destined for the dwarf planet Pluto and the Kuiper Belt.

right
Dragon Storm
A false-color image of Saturn made from Cassini images taken in infrared light show this planet is far from serene. The reddish feature above center is Dragon Storm, a giant thunderstorm.

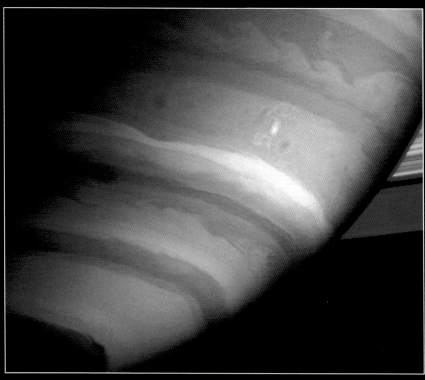

left
Cassini at Saturn
Saturn's rings cast a shadow on to the planet in this March 2007 view taken by the Cassini spacecraft. One of the planet's moons Dione hangs in the distance beyond Saturn.

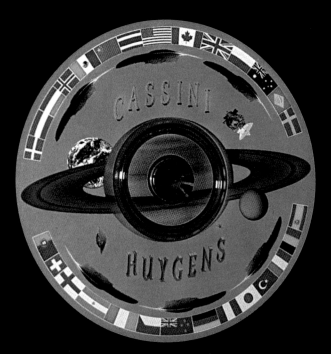

left
Message from Earth
The Cassini spacecraft carries a DVD with 616,400 signatures of people from all around Earth. The disk is decorated with flags from 28 nations. Saturn, Earth, Titan, Cassini, and Huygens are also represented.

Star Voyager

One probe stands out for its outstanding achievements. It is Voyager 2, the first of an identical pair launched in 1977. The prime targets of both were Jupiter and Saturn, and both probes sent back incredible images of the planets, their moons and rings. While Voyager 1 moved off into interplanetary space, Voyager 2 continued to explore other worlds.

The Voyager mission team had decided to make a "grand tour" of the giant planets, taking advantage of a planetary configuration that would not occur again for some 176 years. After picking up gravitational energy as it passed Jupiter (1979) and then Saturn (1981), Voyager 2 went on to encounter Uranus (1986) and Neptune (1989).

Building on success

Before the Voyagers had completed their work, plans were in hand for a new mission to Jupiter. From late 1995 until 2003, Galileo made a thorough study of the Jovian system; orbiting the planet and flying close to its four major moons. In 2004, 23 years after Voyager 2 had left, the Cassini-Huygens probe arrived at Saturn. The smaller probe Huygens descended to the surface of Saturn's moon, Titan as Cassini started its four-year study of the planet.

Meanwhile, Mars Express, Europe's first venture to Mars returns unprecedented views of the red planet, while two robot rovers, Spirit and Opportunity, explore its surface. Venus Express completes its mapping mission and investigation of the Venusian atmosphere, while New Horizons continues its journey to Pluto, to give us our first close-up view of this distant world in 2015.

The Hubble Space Telescope

Among optical telescopes, the Hubble Space Telescope reigns supreme. It can see farther into the universe than any others, and with greater clarity, from its vantage point hundreds of miles above Earth's obscuring atmosphere. It is named for the US astronomer Edwin Hubble, who pioneered detailed study of the galaxies during the first quarter of the twentieth century using the famous 100-inch (2.5-m) Hooker telescope at Mount Wilson Observatory. The HST's mirror is comparable in size, with a diameter of 95 inches (2.4 m). But it is a much more effective light-gatherer, not only because it operates in space, but also because of its supersensitive CCDs and "state of the art" electronics.

Work began on the HST in the late 1970s, but it was not launched until 1990. "First light" revealed image-blurring due to a flawed primary mirror, and the whole project seemed doomed. But a daring recovery-and-repair mission three years later sharpened the HST's vision and imagery to design specifications. Updated systems and instruments installed on subsequent servicing missions have made the HST an astronomical observatory without equal in the study of the universe.

He might have been a world-class heavyweight boxer; he might have been a brilliant lawyer. But instead Edwin Powell Hubble (1889–1953) chose to to become an astronomer.

Born in 1889 in Marshfield, Missouri, Hubble later moved with his family to Chicago, where he gained a degree in mathematics and astronomy at the University. By 1910, he was studying at Oxford University, gaining a bachelor's degree in Jurisprudence two years later. In 1913, back in the US, he opened a law practice in Kentucky.

Investigating Faint Nebulas

However, the lure of the heavens proved too strong for Hubble, and he returned to Chicago to become a graduate student at Yerkes Observatory. In 1917, he gained a doctorate with his thesis "Photographic Investigations of Faint Nebulae," in which he suggested that spiral nebulas might lie outside our own star system.

After a stint with the American Expeditionary Force in France during World War I, Hubble accepted an invitation by George Ellery Hale to work at Mount Wilson Observatory. It was there, in 1919, that he began observing with the newly completed 100-inch (2.5-m) Hooker Telescope, then the biggest in the world. The Hooker still wasn't powerful enough to resolve individual stars in spiral nebulas, although it could detect pinpricks of light in them that some thought might be novas.

In 1923, Hubble spotted one in a photograph of the Great Spiral in Andromeda. He then checked previous photographic plates and found points of light in the same position but with different brightnesses. He plotted the light curve of the source and found that it had the typical characteristics of a Cepheid variable. Applying Henrietta Leavitt's period-luminosity law, Hubble deduced that the Cepheid was an astonishing 900,000 light-years away.

This meant that the Andromeda Nebula was way beyond the confines of our own Milky Way Galaxy. It had to be a separate star system—a separate galaxy.

At the time the Milky Way was estimated to be 300,000 light-years across—three times its actual size. Hubble's value of the distance to the Great Spiral on the other hand was a gross underestimate; the true distance is about 2.5 million light-years.

Hubble's Tuning Fork

Hubble eventually prepared a paper announcing this momentous discovery which Henry Norris Russell presented on New Year's Day 1925. Shortly after, Hubble introduced a system for classifying the galaxies, or "extragalactic nebulae," as he preferred to call them. There were irregulars, with no particular shape, and regulars, with a characteristic shape. He devised the famous "tuning fork" system, still used, which classifies the regulars by their shape into ellipticals, spirals and barred spirals (see page 70).

left
Edwin Hubble
The foremost US astronomer of modern times, whose work led to an appreciation of the true scale of the universe.

Hubble also continued, by looking for Cepheids, to determine the distance to other galaxies. Meanwhile, at Lowell Observatory in Arizona, astronomer Vesto Slipher was estimating the speeds at which galaxies were moving by examining the spectrum of their light. He found that almost all had a red shift in spectral lines, indicating that they were rushing away from us, and at unheard-of speeds.

An Expanding Universe

By correlating his distance values with Slipher's speeds of recession, Hubble made an astonishing discovery. The farther away a galaxy was, the faster it receded. Hubble announced this startling fact in 1929, and it became known as Hubble's Law. It led to the fundamental concept of modern cosmology — that the universe is expanding.

Just as happened in World War I, World War II interrupted Hubble's work in astronomy. He became chief of ballistics and director of the Supersonic Wind Tunnel Laboratory in Maryland. After the war, Hubble became involved with the 200-inch (5-m) Hale Telescope project at Mount Palomar Observatory. When the telescope became operational in 1949, it was fitting that Hubble should begin observations. With this powerful instrument, he could at last resolve the stars in the Great Spiral and other spirals.

By this time, Hubble had only a few years to live. He had developed a heart condition and in 1953 died from a cerebral thrombosis. It is altogether appropriate that the Hubble Space Telescope, which is every day expanding our knowledge of the universe, is named for the man who proved that the universe itself is expanding.

above
Hubble sunrise
Sunlight turns the HST into liquid gold as the Sun peeps over the limb of the Earth, hundreds of miles away. It was one of the 16 sunrises per day the STS-109 mission astronauts experienced on their servicing mission in March 2002.

Getting Into Orbit

In 1923, the same year that Edwin Hubble proved that spiral nebulas were external galaxies, German rocket scientist Hermann Oberth wrote a groundbreaking book about space travel called (in translation), *The Rocket into Interplanetary Space*. In it he discussed the possibility of a space telescope that could be attached to an orbiting space station, or even mounted on a small asteroid for stability.

Oberth's ideas were far ahead of his time. Not until 1946 did the concept of a space telescope resurface, in a paper written by the US astronomer Lyman Spitzer. It was entitled "Astronomical Advantages of an Extra-Terrestrial Observatory." Only 11 years later, Sputnik 1 went into orbit and space flight became a reality.

The Large Space Telescope

In the 1960s, Spitzer began to lobby NASA and the US Congress to provide funding for a large orbiting telescope. By the decade's end, NASA's first orbiting telescope (OAO-2) was operational, and a committee of the National Academy of Sciences had published a report on "Scientific Uses of the Large Space Telescope." Funding for a proposed 120-inch (3-m) instrument was elusive, and so plans were drawn up for a cheaper 95-inch (2.4-m) version.

Congress eventually approved funding for the smaller instrument, and in 1977 NASA began work on the project, known as the Large Space Telescope (LST). By then, NASA had acquired a partner, the European Space Agency (ESA). Under their agreement, ESA would provide 15 percent of the hardware, in return for 15 percent of observing time. The launch was scheduled for 1983.

Lift-off

By 1979, the detailed design and specifications of the LST had been finalized, and construction began. With the involvement of more than 20 major contractors, a university and three NASA centers, building the LST was far from straightforward. It also proved much more expensive that anticipated.

In the event, the telescope was nowhere near ready for launch in 1983, when it was officially named the Edwin P. Hubble Space Telescope. A possible 1986 launch was delayed by the grounding of the shuttle fleet following the Challenger disaster. Even when the shuttle fleet became operational again in 1988, the HST had to wait in line. Not until April 24, 1990, did the telescope ascend into the heavens. The Kennedy Space Center's launch commentator announced: "Lift-off of the space shuttle Discovery with the Hubble Space Telescope — our window on the universe."

How the HST Works

The HST orbits the Earth about once every 95 minutes, in a nearly circular path about 350 miles (560 km) high. This height is well within reach of the shuttles that service it. However, this is a relatively low altitude for a satellite, and its orbit gradually decays because of faint traces of atmospheric gases still present. So, after a service mission, visiting shuttles redeploy it in a slightly higher orbit.

The HST is a big satellite, over 43 feet (13 m) long, with a diameter of 14 feet (4.3 m). It weighs about 12.5 tons (11.5 tonnes). Designed for regular servicing while in orbit, it is outfitted with hand-holds and also grapple fixtures so that it can be gripped by the shuttle orbiter's RMS (robot) arm. Its instruments are modular, so that each can be repaired or replaced without affecting the others.

Two large solar arrays, made up of thousands of solar cells, provide the HST with electricity to keep its batteries charged. The latest panels can supply more than five kilowatts. Unlike the original panels, which were flexible and unfurled, the present ones are rigid. They are also smaller, measuring about 23 feet (7.1 m) long and 8.5 feet (2.6 m) wide.

above

The HST in orbit

The main body tube houses a fairly conventional reflecting telescope, which feeds focused light into instruments in the aft assembly. Data gathered by the telescope is transmitted back to the ground by radio from a pair of antennas. Twin panels (arrays) of solar cells provide the necessary electrical power.

left

Hands-on

STS-82 astronauts Steven Smith (left) and Mark Lee gather their tools in preparation for another day's work on a routine servicing mission. The HST was the first major satellite designed specifically for in-orbit servicing.

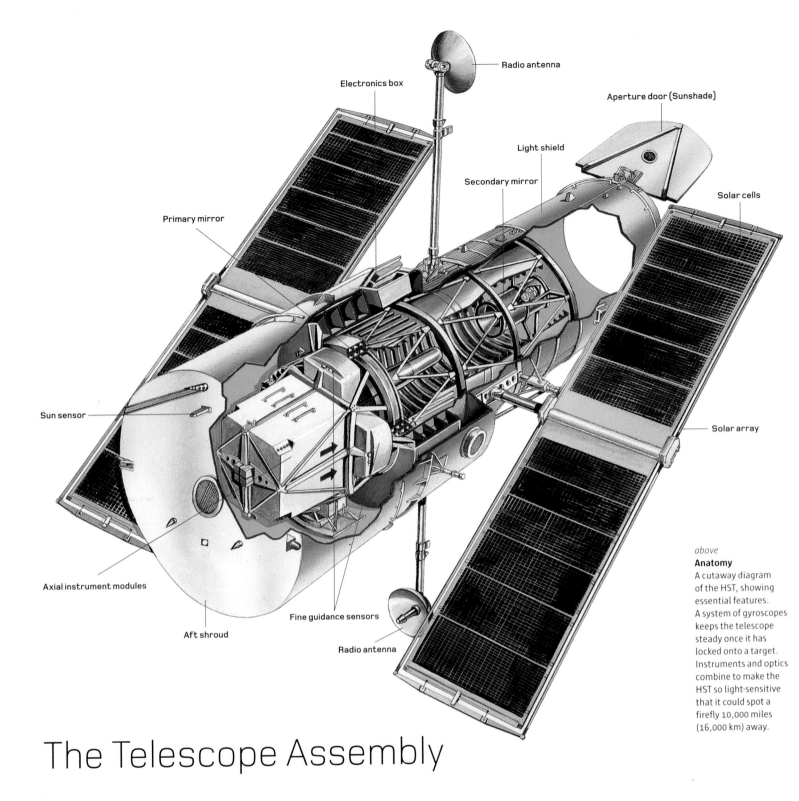

Radio antenna

Electronics box

Aperture door (Sunshade)

Light shield

Secondary mirror

Solar cells

Primary mirror

Sun sensor

Solar array

Axial instrument modules

Fine guidance sensors

Aft shroud

Radio antenna

above

Anatomy

A cutaway diagram of the HST, showing essential features. A system of gyroscopes keeps the telescope steady once it has locked onto a target. Instruments and optics combine to make the HST so light-sensitive that it could spot a firefly 10,000 miles (16,000 km) away.

The Telescope Assembly

The optical design of the HST literally mirrors that of Earth-based telescopes. It is a reflecting telescope that uses mirrors to gather and focus light.

The telescope is a Richey-Chrétien variant of a standard Cassegrain design. In a Cassegrain telescope, light gathered by the large primary mirror is reflected up to a smaller secondary mirror. Then the secondary reflects the light back down the telescope tube and through a hole in the center of the primary, bringing it to a focus underneath. In the case of the HST, the primary mirror is 95 inches (2.4 m) in diameter.

The Science Instruments

Light focused by the optical system feeds into a variety of cameras and other science instruments, located behind the primary mirror in the aft shroud. The five original instruments on the HST were the High-Speed Photometer, the Wide-Field and Planetary Camera, the Faint Object Camera, the Goddard High-Resolution Spectrograph and the Faint Object Spectrograph. All these instruments were replaced during the first four servicing missions.

Wide-Field and Planetary Camera (WF/PC)

The WF/PC is the most widely used of the instruments on the HST. It comprises two cameras, each using CCDs (charge-coupled devices) to record images. In wide-field mode, it looks at large areas of sky, while the planetary camera provides higher resolution and magnification. The original instrument, WF/PC-1, was removed on the first servicing mission and replaced by WF/PC-2, which contains a system for correcting the flawed optics of the primary mirror. When data from all four CCDs in the instrument are displayed together, it produces a chevron-shaped edge, because the scale of the planetary camera image needs to be reduced to that of the wide-field camera.

Faint Object Camera (FOC)

As its name implies, the FOC was designed to see faint objects in the greatest detail. It used an electronic light detector called an image intensifier. It was so sensitive that, to look at anything brighter than magnitude 21, filters were needed to dim the light, to prevent the detectors from becoming saturated.

The FOC was one of the instruments that COSTAR brought back to full operating capability. After impeccable service, it was replaced on the STS-109 servicing mission in March 2002, by the Advanced Camera for Surveys (see page 176).

High-Speed Photometer (HSP)

The HSP was designed to measure high-speed fluctuations of light from sources of high energy. It was affected by the flawed optics of the primary mirror and the vibrations caused by one of the solar panels. It was removed on the first servicing mission in December 1993 and replaced with the COSTAR assembly, which helped correct the HST's vision.

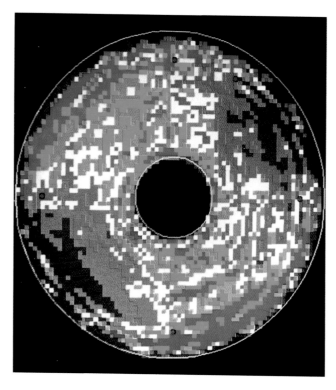

left
Fine limits
Computer scans were used to verify the accuracy of the HST's mirrors, enabling them to be ground and polished to fine limits.

Spectrographs

Spectrographs are designed to produce spectra—smudges of light spread out into separate wavelengths. The dark lines in a spectrum tell astronomers a great deal about the star (or galaxy) that produced the light. Different elements in the star produce different sets of dark lines. A shift in the lines from their normal position indicates that the star may be moving toward us (blue shift) or away from us (red shift), and so on.

The original Goddard High-Resolution Spectrograph (GHRS) and Faint Object Spectrograph (FOS) did excellent work after the COSTAR repair. Both were designed to work in the ultraviolet. On the second servicing mission, in February 1997, they were replaced by the Space Telescope Imaging Spectrograph (STIS), which spans visible and some infrared wavelengths as well as ultraviolet rays.

Fine Guidance Sensors (FGSs)

The HST can only function properly if it is directed to a particular part of the heavens and then kept absolutely still. It therefore has three FGSs, which lock onto particular guide stars and can keep the telescope absolutely steady for long periods.

The FGSs also function as the HST's sixth science instrument, for they are able to provide star positions with at least 10 times the accuracy astronomers achieve on the ground. Measuring star positions is called astrometry.

left
Mirror, mirror
Masked engineers examine the completed primary mirror of the HST in a dust-free cleanroom. The mirror is made from a special silica/titanium oxide glass with minimal thermal expansion and is one foot (30 cm) thick. It took two years to polish.

HST Images

H e might have been a world-class heavyweight boxer; he might have been a brilliant lawyer. But instead Edwin Powell Hubble (1889–1953) chose to to become an astronomer.

Born in 1889 in Marshfield, Missouri, Hubble later moved with his family to Chicago, where he gained a degree in mathematics and astronomy at the University. By 1910, he was studying at Oxford University, gaining a bachelor's degree in Jurisprudence two years later. In 1913, back in the US, he opened a law practice in Kentucky.

Investigating Faint Nebulas

However, the lure of the heavens proved too strong for Hubble, and he returned to Chicago to become a graduate student at Yerkes Observatory. In 1917, he gained a doctorate with his thesis "Photographic Investigations of Faint Nebulae," in which he suggested that spiral nebulas might lie outside our own star system.

After a stint with the American Expeditionary Force in France during World War I, Hubble accepted an invitation by George Ellery Hale to work at Mount Wilson Observatory. It was there, in 1919, that he began observing with the newly completed 100-inch (2.5-m) Hooker Telescope, then the biggest in the world. The Hooker still wasn't powerful enough to resolve individual stars in spiral nebulas, although it could detect pinpricks of light in them that some thought might be novas.

In 1923, Hubble spotted one in a photograph of the Great Spiral in Andromeda. He then checked previous photographic plates and found points of light in the same position but with different brightnesses. He plotted the light curve of the source

and found that it had the typical characteristics of a Cepheid variable. Applying Henrietta Leavitt's period-luminosity law, Hubble deduced that the Cepheid was an astonishing 900,000 light-years away.

This meant that the Andromeda Nebula was way beyond the confines of our own Milky Way Galaxy. It had to be a separate star system—a separate galaxy.

At the time the Milky Way was estimated to be 300,000 light-years across—three times its actual size. Hubble's value of the distance to the Great Spiral on the other hand was a gross underestimate; the true distance is about 2.5 million light-years.

above
Data flow
The circuitous route by which HST data reach the user. Observing time on the HST is at a premium, with on average less than half of interested applicants being successful. But some time is still found for amateur astronomers.

left
WF/PC imaging
HST wide-field/ planetary-camera (WF/PC) views of a barred-spiral galaxy (NGC 1808), a "starburst" galaxy where vigorous star formation is taking place. In the left-hand frame, a combined WF/PC image is superimposed over a ground-based telescope picture. The smaller square shows the planetary camera view shrunk down to the same scale as the wide-field view. This is what causes the chevron edge in many HST images. The more detailed magnified planetary-camera image is shown on the right.

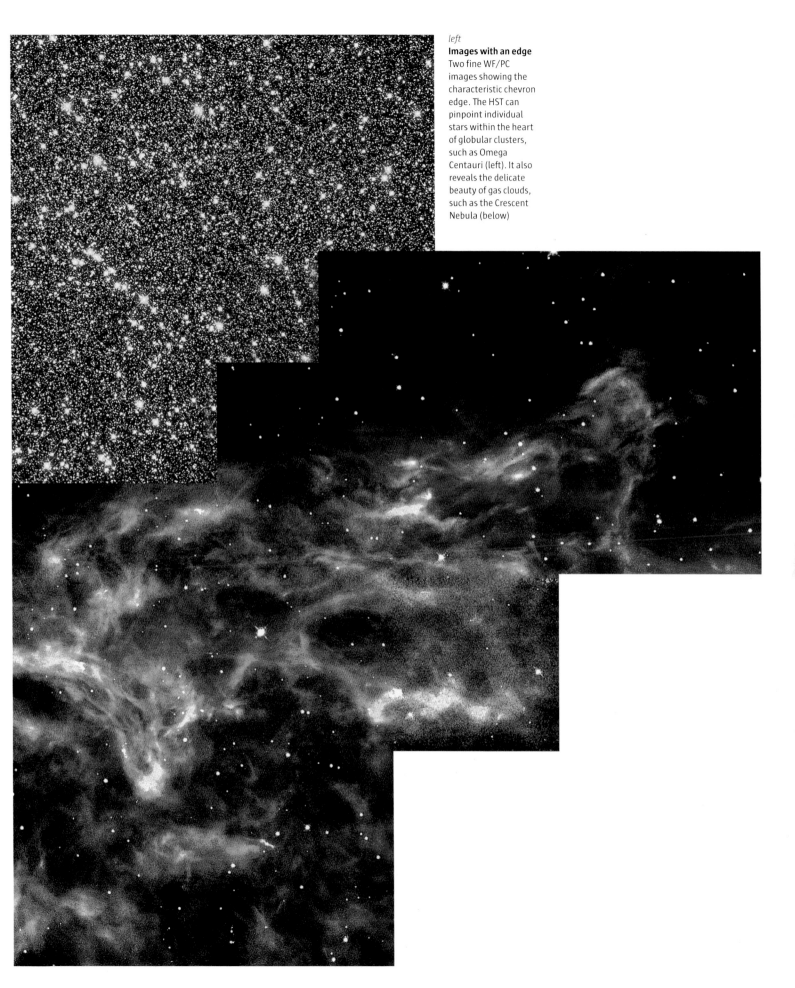

left
Images with an edge
Two fine WF/PC images showing the characteristic chevron edge. The HST can pinpoint individual stars within the heart of globular clusters, such as Omega Centauri (left). It also reveals the delicate beauty of gas clouds, such as the Crescent Nebula (below)

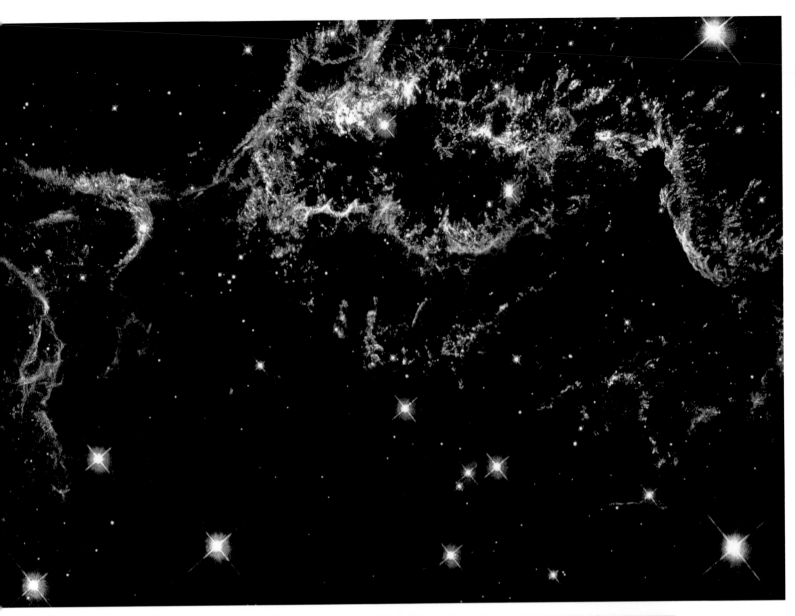

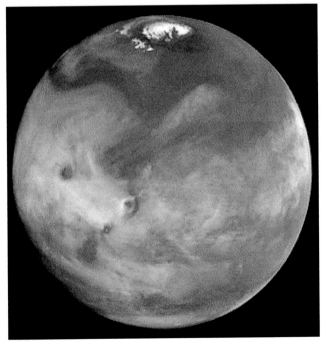

left
Simulated color
The color pictures the HST takes—say, of Mars—are not snapped in color, as we snap pictures on color film. They are snapped at a single wavelength, selected by passing incoming light through a color filter. Reproduced in this form, they present a monochrome, or black-and-white image (left). A true-color image (right) can be simulated by combining a number of images taken through different color filters and assigning them appropriate colors.

Colorful conclusion
Delicate filaments of glowing gas in Cassiopeia mark the remains of a star that blew itself to bits in a supernova explosion that astronomers witnessed in 1572. The HST team has combined and colored images taken through the filters to highlight the presence of different chemical elements. For example, deep blue shows regions rich in oxygen; red, regions rich in sulfur.

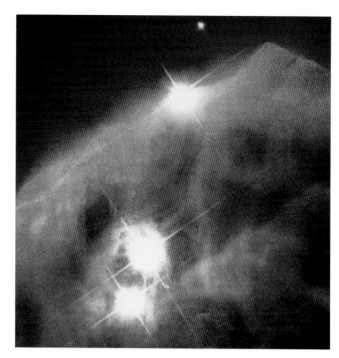

above

Seeing the invisible
These two different views of the same region of the Cone Nebula show how different wavelengths tease out additional details. At left is a visible-light image from the Advanced Camera for Surveys; at right, an infrared image from NICMOS. Note that infrared light has penetrated the dusty haze to reveal a clutch of hot stars.

left

Remote cepheids
This remote galaxy is NGC 4603, just one in a cluster of galaxies in Centaurus. The HST has been able to pinpoint Cepheid variables in the galaxy, establishing its distance at 108 million light-years.

A Flawed Debut

Seven years after its original planned launch date, the HST arrived at the Kennedy Space Center in October 1989. It was installed in the payload bay of shuttle orbiter Discovery on March 29, 1990, with launch planned for April 10. The HST launch mission was designated STS-31.

However, problems with the orbiter's APUs (auxiliary power units) caused the launch to be scrubbed at the T-4 minutes hold in the countdown. Two weeks later, on April 24, the countdown went all the way , and Discovery and the HST thundered into the heavens.

Next day, with Steve Hawley at the controls, Discovery's robot arm picked up the HST and lifted it out of the payload bay. When Hawley sent commands to unfurl the twin solar panels, one unfurled but the other jammed. Inside Discovery's airlock, astronauts Kathy Sullivan and Bruce McCandless were suited up, ready to go on EVA to free the jammed solar panel manually. But eventually it freed itself.

Hawley released the robot arm's hold on the HST, which stayed close by. Two days later, it took two attempts to open the aperture door of the HST. It was at this point that Story Musgrave, the CapCom (capsule communicator) at Mission Control, radioed: "Discovery, Hubble is open for business."

First Light

As it happened, the HST wasn't quite "open." The first few weeks after launch were used for equipment, systems and engineering checks. Not until May 20, 1990, did the HST experience "first light"—this is the time when a telescope becomes fully operational and returns the first image.

This long-anticipated event unfolded before the eyes of journalists from around the world at the Space Telescope Science Institute, who cheered as an image flashed onto the screen before them. The portrait of the open star cluster NGC 3532 was actually a somewhat boring image, but it was greeted with elation by the press and with more than a little relief by mission scientists. This was the first evidence that the HST actually worked!

Behind the scenes, however, HST scientists were worried. They realized that the first-light image was blurred. And no matter what they did, they couldn't make it sharp. The HST wasn't focusing properly! Word soon leaked to the once adulatory press, who were quick to pour scorn: "Pix nixed as Hubble sees double" screamed one memorable headline.

The HST team soon realized that their US$1.5 billion brainchild had flawed optics. The primary light-gathering mirror was suffering from a classic curved-mirror defect known as spherical aberration. The mirror had an incorrect curvature, which caused light rays from different parts of the mirror to be brought to a focus at different points.

It turned out that the primary mirror deviated from the correct curvature by a miniscule amount—just two microns, or about one-fiftieth of the width of a human hair. Still, this was enough to blur the images. The defect had apparently been caused in manufacturing and had not been detected during subsequent tests.

left
Smile please
Discovery's crew poses for the traditional on-board photograph. The astronauts are Charles Bolden (at top), then, from left to right: Loren Shriver, Kathryn Sullivan, Bruce McCandless and Steven Hawley.

right
Into orbit
On April 25, 1990, Discovery's robot arm plucks the HST from the payload bay and places it in orbit. It is ready for business, or so everyone thought.

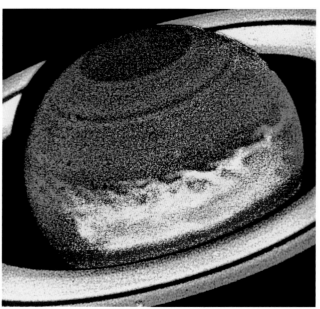

far left
Lift-off!
All engines blazing, shuttle orbiter Discovery thunders away from the launch pad at the Kennedy Space Center carrying the HST. It is April 24, 1990. The HST is late—very late.

left
Stormy giant
In the fall of 1990, astronomers reported seeing a white spot on Saturn, marking the site of an exceptional storm. This is one of the time-lapse images the HST took to track how the storm developed; it has been color-enhanced.

Emergency Service

So here was a telescope hundreds of miles away in space that was returning blurred pictures due to faulty optics. While NASA and the Space Telescope Science Institute debated what to do, computer experts were devising a way to manipulate the incoming data in order to improve the quality of the images. This computerized image-processing technique, called deconvolution, involved determining where light was missing from the image and restoring it.

Deconvolution and longer exposure times certainly helped to produce acceptable images, but they were still only on a par with the best images taken by the most powerful observatories on Earth under good conditions. The mammoth leap in observational astronomy that scientists were expecting from the HST had not materialized.

How to fix the HST—that was the problem. Some suggested that it should be recovered, returned to Earth, refurbished here and then redeployed. But the consensus was that the HST should be repaired in orbit. After all, it been designed for in-orbit servicing.

COSTAR to the Rescue

The Hubble team soon devised a means of correcting the telescope's defects. They would use an ingenious piece of equipment with 10 small mirrors, called COSTAR (corrective optics space telescope axial replacement). COSTAR would clarify the Hubble's vision rather like eyeglasses cure defective eyesight. Its mirrors would refocus the light coming from the defective primary and feed it to the Faint Object Camera (FOC) and spectrographs. However, in order to make room for COSTAR aboard the HST, the High-Speed Photometer (HSP) would have to be removed.

The HST team would also replace the existing Wide Field and Planetary Camera (WF/PC-1) with a "clone" (WF/PC-2) that had reconfigured mirrors. These, too, would combat the effects of the mirror's aberration.

Mission Accomplished

By December 1993, NASA was ready to launch a rescue and repair mission, designated STS-61. And just in time. By now, the HST really was in trouble and on the verge of becoming useless. Three of its six gyroscopes had failed, and so had two of its memory banks. In addition, its solar panels vibrated every time it passed between day and night or light and shade, which happened 16 times every 24 hours.

Just before dawn on December 2, 1993, shuttle orbiter Endeavour blasted off on the vital first servicing mission,. It rendezvoused with the HST two days later. Using the robot arm, Endeavour captured the telescope and berthed it in the payload bay. On December 5, veteran spacewalkers Story Musgrave and Jeff Hoffman made the first of five EVAs (extravehicular activities) to make repairs. It took them nearly eight hours to replace the faulty gyroscopes and other electronic parts.

left
Capture
It is December 4, 1993. Shuttle orbiter Endeavour maneuvers closer and closer to the ailing HST, seen here against a backdrop of the Indian Ocean off Australia's west coast. Soon the HST will be safely berthed in Endeavour's payload bay.

left
New for old
On their EVA of December 7, Jeff Hoffman and Story Musgrave install the new WF/PC-2. Here Hoffman is seen with the old instrument, which he will stow in the payload bay.

On December 6, Kathryn Thornton and Tom Akers went on EVA to replace the two solar panels. Musgrave and Hoffman returned the next day to install the new WF/PC. On December 8, Thornton and Akers removed the High-Speed Photometer (HSP) and fitted COSTAR into place. Musgrave and Hoffmann carried out the last of the five spacewalks later the same day, bringing the total EVA time for the mission to more than 35 hours—a shuttle record. On December 9, Endeavour's robot arm lifted the HST out of the payload bay and back into independent orbit. Endeavour itself returned to base four days later.

Second First Light

From NASA's point of view, the STS-61 servicing mission had gone amazingly well. But it remained to be seen whether the repair team's "eye surgery on the patient" had corrected the telescope's faulty vision. On December 18, 1993, the HST science team gathered to witness first light of the refurbished instrument. An image of a star, Melnick 34 in the 30 Doradus (Tarantula) Nebula, appeared on the screen as a bright point of light—not spread out or blurred. The repair mission had worked.

By January 13, 1994, the HST team had acquired a portfolio of spectacular images taken with the rejuvenated telescope, which they shared with the press. "The patient," an HST spokesman said, "has a new vision of incredible clarity."

New Vision

Over the next four years, the HST returned spectacular images by the hundreds, proving that it had indeed opened up a new

above
Nearly done
On the final EVA of the STS-61 repair mission, Story Musgrave hitches a ride on the robot arm to fit protective covers over the magnetometers, instruments designed to detect magnetic fields.

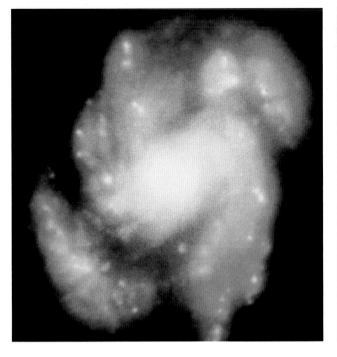

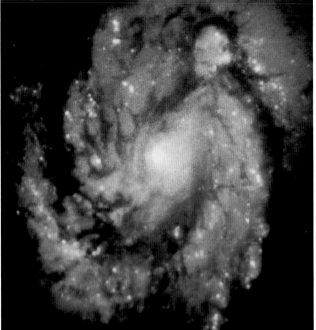

left
New vision
Two HST images that tell the story of the success of the first make-or-break service mission. The image of the spiral galaxy M100 before the mission (left) is little better than views from ground telescopes. The image of the galaxy after the repair is infinitely clearer.

window on the universe. In February 1997, a planned second servicing mission (STS-82) was underway. Discovery lifted off with a number of new instruments to improve the telescope's visual acuity and range.

Spacewalking astronauts, working alternately in pairs—first Mark Lee and Steve Smith and then Greg Harbaugh and Joe Tanner—carried out five EVAs to install the new equipment. On the first and most important EVA, Lee and Smith removed the two existing spectrographs and replaced them with the Space Telescope Imaging Spectrograph (STIS) and the Near-Infrared Camera and Multi-Object Spectrometer (NICMOS). The STIS would provide improved spectral resolution, while NICMOS would extend the sight of the HST into near-infrared wavelengths.

Subsequent EVAs saw the replacement of a number of instruments and equipment, including one of the Fine Guidance Sensors (FGSs), a data recorder and a Reaction Wheel Assembly (RWA). On the final EVA, astronauts Lee and Smith repaired damaged thermal insulation on the outside of the HST, which protects the internal structure and instruments from fluctuations in temperature.

Into Hibernation

A third HST servicing mission was planned for June 2000, but one after another, its attitude-maintaining gyroscopes began to fail in 1997, 1998 and 1999. So the HST team decided to split the objectives of the mission in two, with the first launch in December 1999. On November 17 the fourth gyro failed, leaving just two gyros functioning—not enough to position the instrument. The telescope had to be shut down—it was put into "hibernation mode."

Discovery blasted off on the STS-103 service mission (designated SM3A) on December 19, 1999. On three spacewalks, astronauts replaced the faulty gyros, fitted kits to improve the telescope's batteries and installed a new computer that was 20 times faster and had six times as much memory as the original.

Advanced Imaging

Shuttle orbiter Columbia sped into orbit on March 1, 2002, on the STS-109 service mission (designated SM3B). The crew kept a rigorous schedule in order to install all of the new and upgraded instruments and equipment.

The critical new instrument was the Advanced Camera for Surveys (ACS), which replaced the Faint Object Camera (FOC). The FOC had been "state of the art" in the 1980s, but digital imaging had improved dramatically since then.

The ACS comprises three separate instruments: the High-Resolution Channel, the Solar-Blind Channel and the Wide-Field Channel. The High-Resolution Channel takes detailed

left
Instrument update
Astronauts move in to install new equipment in the HST on the STS-82 mission in February 1997. Joe Tanner stands on the platform on the robot arm, with Greg Harbaugh nearby. On five EVAs during the mission, astronauts notched up a total time of more than 33 hours.

right
The Pistol
The newly installed NICMOS instrument gave the HST the ability to see in the near-infrared. It was soon returning spectacular images—which of course had to reproduced in false-color. This image centers on the Pistol Star and the surrounding nebula. It is one of the most luminous stars we know.

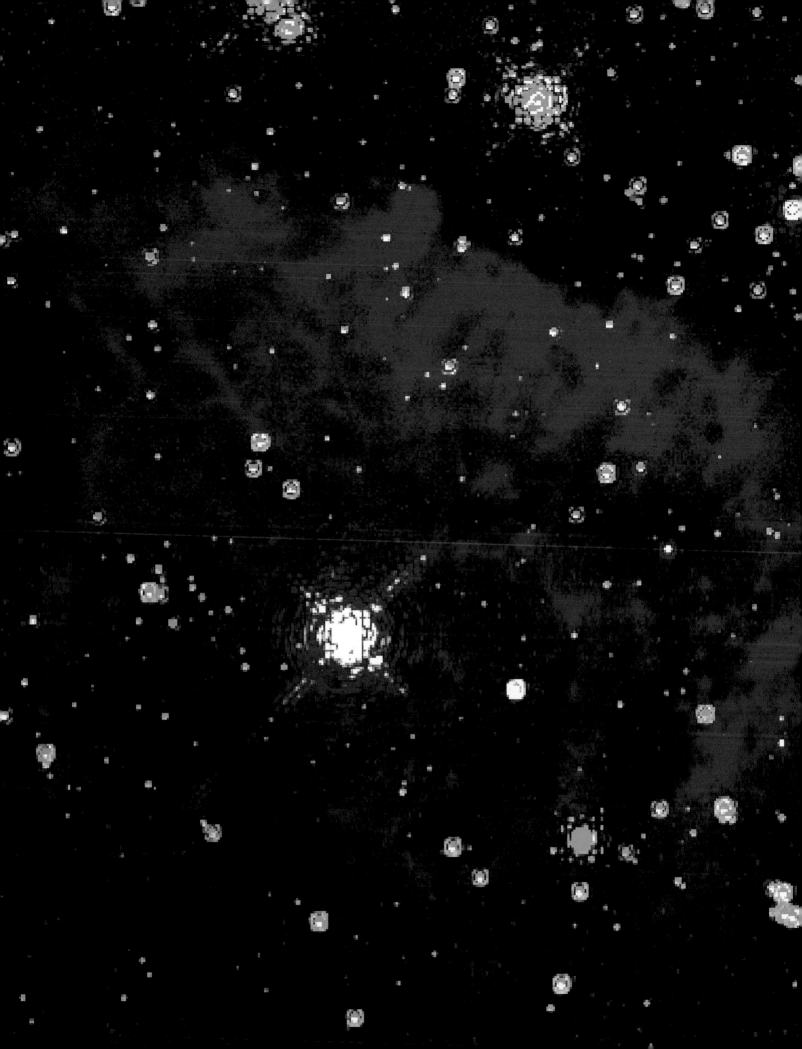

left
End of the arm
Richard Linnehan is anchored on the end of shuttle's robotic arm. He is unfolding a solar array during the STS-109 service mission in March 2002.

below
Eruptive star
The Advanced Camera for Surveys (ACS) installed on the STS-109 mission works brilliantly. This ACS image shows the unusual variable star V838 Monocerotis. The HST has returned to it many times since it suddenly erupted in 2002.

images of the inner regions of galaxies and searches for extrasolar planets. The Solar-Blind Channel, which blocks visible light to enhance ultraviolet sensitivity, studies planetary weather in our own solar system. The Wide-Field Channel helps astronomers study the nature and distribution of the galaxies on a broad scale.

The ACS was installed on the fourth of the five EVAs of the mission. Earlier, astronauts had replaced the solar arrays with a smaller but more powerful set and installed a new Reaction Wheel Assembly (RWA).

On the fifth EVA, a new neon cryocooler — a refrigeration unit — was fitted for the NICMOS. Working in the infrared, its instruments must be cooled to around minus 330 degrees Fahrenheit (–200°C). The original nitrogen-ice coolant had run out in January 1999, earlier than expected. Installation of the new cooler would bring the instrument back into operation.

The five EVAs on STS-109 broke the duration record for a single flight, totaling nearly 36 hours. Over the four HST servicing missions to date, 14 astronauts had spent a total of more than 129 hours on maintenance work.

Last mission

A further servicing mission (SM4) was planned for 2004 but was postponed, then cancelled, after the Columbia space shuttle tragedy in 2003. It is now reinstated and scheduled for late 2008. This will be the fifth and final mission to Hubble and it will ensure the HST's survival until at least 2013 when its successor will be ready to take over.

The servicing mission will be a 7-astronaut, 11-day space shuttle flight and involve a number of spacewalks. Two new instruments will be installed. These are the Cosmic Origins Spectrograph (COS), an ultraviolet spectrograph which will probe the large-scale structure of the universe, and the Wide Field Camera 3 (WFC3), which will continue the HST's work on the solar system, star formation, and galaxies.

An attempt will also be made to repair the Space Telescope Imaging Spectrograph (STIS) which was installed in 1997 and stopped working in mid-2004. Other work will include the installation of a refurbished Fine Guidance Sensor to replace one of the three that currently control the telescope's pointing system, and new batteries and thermal blankets will be put in position. It is also possible that some work will be carried out on the Advanced Camera for Surveys. This was installed on a previous servicing mission in 2002 but stopped working in January 2007.

Toward the Future

The HST's successor is the James Webb Space Telescope (JWST) named for James Webb, NASA administrator during the Apollo era. The JWST, which will work in the visible to mid-infrared wavelengths will not, like the HST, be positioned in orbit round Earth. It will work from a gravitationally stable location known as Lagrangian Point 2 some 940,000 miles (1,500,000 km) away, beyond Earth's Moon.

Work began on the eighteen segments of the primary mirror in 2004. When complete, in 2009, the mirror will measure 21.3 feet (6.5 m) across. Other telescope components are under construction and the instrument will be assembled and tested in 2012 ready for launch by the European rocket Ariane 5 in 2013. The JWST is expected to last for a minimum of five, and up to ten years.

The mirror will be folded for launch but once in orbit will unfurl like a butterfly opening its wings. A sunshield as big as a tennis court will also deploy after launch to protect the telescope from the Sun's heat and light. The mirror is about six times the area of the HST's enabling the JWST to see further back in time than the HST, to about 300 million years after the Big Bang.

Astronomers hope that the JWST will shed much more light on the "dark ages of the universe," the period when the first stars and galaxies were forming. It will observe the building blocks of galaxies, and examine the birth and evolution of stars. How this new space telescope will shape up remains to be seen, but what is clear is that the HST is going to be a very hard act to follow.

below
New wings
With smaller but more powerful solar arrays, NICMOS back in working order, and a new and more advanced camera, the HST begins to orbit independently at the end of the STS-109 mission.

following pages
Hubble's heritage
The HST will long be remembered for stunning images. The bluish-white dots scattered throughout this view of galaxy Arp 220 are star clusters. The clusters are so compact they look like single stars.

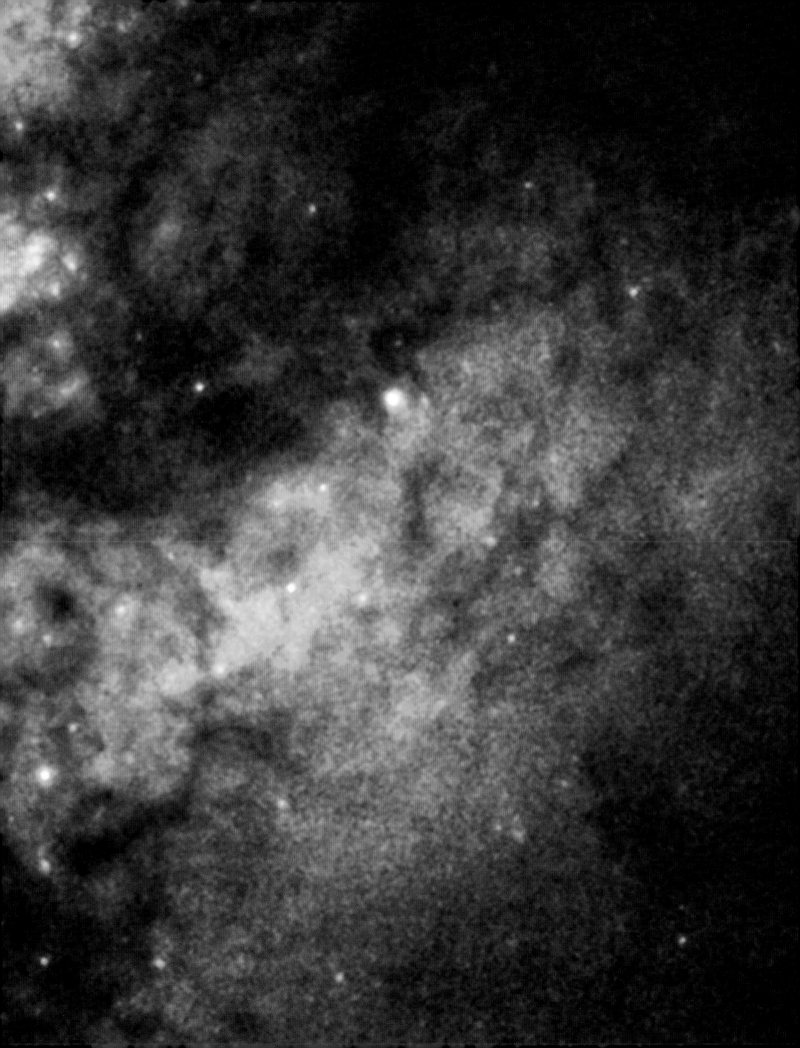

Glossary of Terms

absolute magnitude The true brightness of a star; the magnitude that would be observed from a distance of 10 parsecs (33 light-years)

accretion disk A disk of gas and dust that forms around newborn stars and black holes

active galaxy A galaxy with an exceptionally high energy output (often as radio waves) thought to originate from a massive black hole

active optics A technique for building large telescope mirrors. It uses a number of mirror segments that are aligned together into a perfect shape by computer-controlled actuators

antimatter Matter made up of particles that are the exact opposite of ordinary particles, such as positrons (antielectrons, with a positive electric charge) and antiprotons (with a negative charge)

apparent magnitude The brightness of a star as we see it in the sky

asterism A close grouping of stars in the sky

asteroid A chunk of rock or metal circling in space, mainly in a band (the asteroid belt) roughly mid-way between the orbits of Mars and Jupiter

astrology An ancient belief that human lives and fortunes are somehow affected by the relative positions of the Sun and the planets among the stars

astronomical unit (AU) The average distance between the Earth and the Sun—93,000,000 miles (150,000,000 km)

astronomy The study of the heavens and heavenly bodies

atmosphere The layer of gases that surrounds the Earth and other planets; a unit of pressure, being the pressure exerted by the Earth's atmosphere at sea level

atom The smallest unit of matter, with a central concentration of mass (called the nucleus) circled by electrons

aurora A glow in the atmosphere of a planet's polar regions due to charged particles interacting with molecules in the upper air. On Earth, in the Northern Hemisphere, the light displays are called the aurora borealis, or Northern Lights, and in the Southern Hemisphere the aurora australis, or Southern Lights. The HST has spotted auroras on Jupiter and Saturn

barred-spiral A kind of spiral galaxy that has a band of stars extending through its nucleus

Big Bang The generally accepted theory that the universe was created in a fantastic explosion roughly 12 to 15 billion years ago

Big Crunch The theory that the universe will end with all matter coming together and being crushed to nothing; the opposite of the **Big Bang**

billion One thousand million: 1,000,000,000

binary A double-star system. Two stars, bound by gravity, orbit their common center of mass

black hole A region of space that has such powerful gravity that not even light can escape. One can form when a high-mass star collapses or when a very large mass collapses in a galaxy's core

blazar An active galaxy with radiation that seems (to us on Earth) to be coming straight from its core

blue shift A shift in the lines in the spectrum of a star toward the blue end, indicating that the star is approaching us

brightness *See* **absolute magnitude**, **apparent magnitude**

brown dwarf A small, failed star, whose nuclear furnace has never lit up. It produces some heat but little light

CCD Charge-coupled device, an electronic imaging device used in most modern telescopes (including the HST) to acquire images

celestial sphere An imaginary dark sphere that appears to envelop the Earth, on the inside of which the stars are fixed. It appears to revolve round the Earth once a day from east to west

Cepheid A variable star that changes brightness over a regular, short period; used to estimate distances to remote galaxies

chromosphere The colorful inner atmosphere of the Sun that becomes visible during a total solar eclipse

cluster A grouping of stars or galaxies; *see also* **globular cluster, open cluster, supercluster**

coma The bright head of a comet

comet An ice, snow, and dust body that develops a coma and one or more tails when near the Sun

constellation A pattern of stars that seem to be grouped together because they appear in the same part of the sky

core The central region of a planet, Sun or star

corona The outer atmosphere of the Sun, visible from Earth only during a total solar eclipse

cosmic ray A fast-moving, charged atomic particle emanating from space

cosmology The study of the origin and evolution of the universe

cosmos An alternative name for the universe

COSTAR Corrective optics space telescope axial replacement; an assembly of mirrors fitted to the HST to correct the flawed vision of its light-gathering primary mirror

crater A circular pit in the surface of a planet or moon, usually made by the impact of a meteorite

crust The solid surface of a planet or moon

dark matter Unseen matter that is thought to make up as much as 90 percent of the mass of the universe

declination Celestial latitude; the distance a star is from the celestial equator, measured in degrees

degenerate matter A kind of highly compressed matter found in white-dwarf and neutron stars

double star A star that looks like a single star but is actually two stars close together. Some doubles are two stars that happen to appear together in our line of sight. Others are pairs of stars that really are close together, called binaries

dwarf planet A term introduced in August 2006 to describe a planetary body that orbits the Sun, is nearly round, has not cleared its neighborhood, and is not a satellite. Eris, Pluto, and Ceres are dwarf planets

eclipse When one heavenly body passes in front of another and blocks out its light; *see also* **lunar eclipse**, **solar eclipse**

eclipsing binary A kind of variable star, which is actually a binary star system. The two stars orbit in our line of sight and periodically eclipse each other. When they do, the overall brightness of the system drops

electromagnetic radiation Waves of energy emitted by stars and other heavenly bodies —gamma rays, X-rays, ultraviolet waves, visible light rays, infrared rays, microwaves and radio waves. They all travel at the same speed (the speed of light) but have different wavelengths

electron A tiny subatomic particle present in all atoms that has a negative electric charge

element A basic building block of matter that cannot be broken down into a simpler chemical substance

elliptical galaxy A galaxy of a round or oval shape that lacks spiral arms

encounter A meeting in space between a spacecraft and a planet or other heavenly body

ESA European Space Agency, the organization that coordinates space activities in Europe

escape velocity The speed at which a body must travel to escape the gravity of another: Earth's escape velociy is about 25,000 miles per hour (40,000 km/h)

EVA Extravehicular activity, work outside a spacecraft; popularly called spacewalking

evening star Either the planet Venus or Mercury, seen in the western sky after sunset

expanding universe The theory that the universe is expanding, evidenced by the recession of the galaxies

extrasolar planet A planet in orbit around a star other than the Sun

extraterrestrial Not of the Earth; an extra-terrestrial being or an alien

falling star A common name for a meteor

false-color image An image that is reproduced in colors that are not those we would see with our eyes

filament A string of galaxy superclusters that extends over a vast distance in space

first light The first time a telescope produces an image

flyby A space mission in which a probe flies past a planet without going into orbit or landing

fusion *See* **nuclear fusion**

galaxy A vast group of stars, gas, and dust, bound by gravity and having a mass greater than 100,000 masses of the Sun. The Milky Way is an example

gamma rays Electromagnetic rays that have the shortest wavelengths and the most energy

gas giant Jupiter, Saturn, Uranus or Neptune, which are all made up mainly of gas and liquid and are giant-sized compared with the terrestrial (Earth-like) planets

globular cluster A huge, globe-shaped mass containing hundreds of thousands of old stars that are grouped closely together

gravitational lens The distortion of the light from a distant object when it passes through a strong gravitational field, such as that of a galaxy or galaxy cluster. It usually results in multiple images of the object

gravity The force of attraction that exists between bodies. The more massive a body is, the stronger its gravitational attraction

greenhouse effect A situation in which the atmosphere traps heat like a greenhouse and causes global warming

halo The spherical region around a galaxy that contains globular clusters and dark matter

heavens The night sky. Hence the term heavenly bodies: the objects we see in the night sky

helium The second-most common element in the universe, after hydrogen; formed in stars by the nuclear fusion of hydrogen

H-R diagram The Hertzsprung-Russell diagram; a graph that correlates the absolute brightness of stars with their spectral class or temperature

Hubble Constant A quantity describing the speed at which galaxies accelerate away from one another and, therefore, how fast the universe is expanding

Hubble's law The velocity of recession of a galaxy is proportional to its distance. The constant of proportionality is the Hubble Constant

hydrogen The element in the universe that is the most common, the simplest (composed of one proton and one electron) and the lightest

ice world One of many bodies found in the outer solar system and composed principally of ice. Eris and Pluto are the two largest

infrared rays Electromagnetic rays that are slightly longer in wavelength than visible red light; heat radiation

intergalactic Between galaxies

interplanetary Between planets

interstellar Between stars

interstellar matter Gas and dust that is scattered in the space between the stars; sometimes visible as clouds or nebulas

irregular galaxy A galaxy with an irregular shape

Kuiper Belt A flattened ring of icy, cometlike bodies that orbit the Sun beyond Neptune. They include Eris and Pluto

lava Molten rock spewed out by volcanoes

light Electromagnetic radiation at wavelengths our eyes can detect

light-year The distance light travels in a year— about 6 trillion miles (9,500,000,000,000 km); used as a unit for measuring distances in space; *See also* **parsec**

LMC The Large Magellanic Cloud

Local Group The small cluster of galaxies to which our own Galaxy belongs. It also includes the Andromeda Galaxy and the Magellanic Clouds

luminosity The absolute magnitude (true brightness) of a star compared with that of the Sun

lunar Relating to the Moon

lunar eclipse An eclipse of the Moon occurs when it passes through Earth's shadow

Magellanic Clouds *See* **LMC, SMC**

magnetic field The region around a star, galaxy or planet where its magnetism acts

magnetosphere A magnetic bubble surrounding a planet that keeps the solar wind at bay

magnitude The brightness of a star; *See* **absolute magnitude**, **apparent magnitude**

main sequence star A star in the prime of its life, located on a diagonal band in the H-R diagram

mantle The rocky layer of a planet between its crust and core

mare A vast, dusty plain on the Moon, named for the Latin word for "sea"; the plural is maria

mass The amount of matter contained in a physical object

matter The "stuff" the universe is made of

meteor The streak of light that appears in the sky when a meteoroid burns up in the Earth's atmosphere

meteorite A lump of rock or metal that falls to the Earth from outer space; the remains of a large meteoroid

meteoroid A piece of rock or metal traveling through space

Milky Way A pale band of stars visible in the night sky that represents a "slice" through our local Milky Way Galaxy

Mira variable A variable star like Mira Ceti that changes brightness over a long period; Mira varies over a period of about 11 months

molecule The smallest part of a substance, made up of a number of similar or different atoms linked together by chemical bonds; *See also* **atom**

molecular cloud An interstellar cloud made up of molecules, such as molecular hydrogen; it is the birthplace of stars

moon The common name for the natural satellite of a planet; the Moon is Earth's only natural satellite

morning star Either the planet Venus or Mercury, seen in the eastern sky before sunrise

NASA National Aeronautics and Space Administration; the organization that coordinates space activities in the United States

near-earth object (NEO) An asteroid that orbits (sometimes dangerously) close to the Earth

nebula A cloud of gas and dust that exists in the space between the stars

NEO *See* **near-earth object**

neutron A subatomic particle with no electric charge, found in the nucleus of every atom except hydrogen

neutron star A very dense star, made up of neutrons packed tightly together. It is a form of degenerate matter

nova A star that suddenly flares up brightly, named for the Latin word for "new"; *See also* **supernova**

nuclear fusion A nuclear reaction in which atoms of a light element (such as hydrogen) fuse, releasing enormous amounts of energy

nuclear reaction A process of change involving the nuclei of atoms

nucleus The center of an atom or a galaxy; also the small, solid part of a comet; the plural is "nuclei"

observatory The place where astronomers work and make their observations

Oort cloud A huge spherical cloud of icy cometary bodies that more or less marks the edge of the solar system

open cluster A loose grouping of up to a few thousand stars

orbit The path in space one body follows when it circles around another. Most orbits are elliptical (oval) rather than circular

orbital velocity The minimum speed a body must have to remain in orbit around another. For a satellite circling a few hundred miles above the Earth, the orbital velocity is around 17,500 mph (28,000 km/h)

parallax The shift in position of a nearby object against a distant background when viewed from different points; used by astronomers to determine the distances to a few hundred of the nearest stars

parsec A unit used by astronomers to measure distances in space; equal to about 3.3 light-years

period A length of time in a repeating cycle; for example, in a Cepheid, the time between peaks of brightness

period-luminosity law The period of variation of a Cepheid variable is directly related to its absolute magnitude, or luminosity

phases Apparent changes in the shape of the Moon or other heavenly body (like Venus), due to more or less of its surface being lit by the Sun

photon A particle of electromagnetic radiation

photosphere The bright, visible surface of the Sun or other star

planet One of the eight main bodies that circle in space around the Sun, including Earth. Many other stars have planets too, known as extrasolar planets. See also **dwarf planet**

planetary nebula A cloud of gas and dust given off by a dying star with a mass similar to that of the Sun

planetesimal A small body formed early in the life of the solar system. Planetesimals congregated together to form the planets

Population I stars Relatively young, hot stars that are typical of spiral arms and contain an abundance of metals, such as O and B stars, supergiants and Cepheids

Population II stars Relatively old red and yellow stars that are typical of the galactic bulge (nucleus) and globular clusters

probe A spacecraft that travels deep into space to explore heavenly bodies

prominence A fountain-like eruption of incandescent gas on the Sun that often follows the invisible loops of the Sun's magnetic field

proplyd Short for "protoplanetary disk"; a disk of matter around a newborn star from which planets might eventually form

proton A subatomic particle found in the nucleus of all atoms. The simplest element, hydrogen, has a single proton in the nucleus

protostar An early stage in the birth of a star, before nuclear processes have begun

pulsar A rapidly rotating neutron star that flashes pulses of radiation toward us, like a celestial lighthouse

quasar Short for "quasi-stellar"; a kind of active galaxy that emits energy from a small, central region. It appears star-like but is more remote than stars and is as bright as many galaxies

radar The technique of bouncing radio waves off objects used in astronomy to map the surface of a planet

radiant The point in space from which the meteors in a meteor shower appear to originate

radio astronomy The branch of astronomy concerning the study of radio waves from space reaching Earth

radio waves Electromagnetic radiation with the longest wavelengths

red giant A large, red star that is near the end of its life. The Sun will swell up 20 to 30 times to become a red giant in about 5 billion years' time

red shift A shift of the dark lines in the spectrum of a star or galaxy toward the red end, indicating that it is moving away from Earth

reflector A reflecting telescope, which uses mirrors to gather and focus incoming light

refractor A refracting telescope, which uses lenses to gather and focus incoming light

resolving power The ability of a telescope to distinguish fine detail

retrograde motion Motion in the opposite direction from normal. Some of Jupiter's moons have retrograde motion, circling the planet in the opposite direction from the others.

right ascension Celestial longitude, measured in hours of sidereal time–time according to the stars

satellite A small body that travels in orbit around a larger one; a moon. It is also the usual term for an artificial, man-made satellite.

sea A flat, dusty plain on the Moon; *See also* **mare**

seasons Regular changes in the temperature and weather caused by the tilt of the Earth's axis in space. Other planets have seasons too, most noticeably Mars.

seeing The quality of the image produced in a telescope, which reflects the clarity and steadiness of the atmosphere

SETI The search for extraterrestrial intelligence

Seyfert galaxy An active galaxy with an exceptionally bright center

shepherd moon A moon that orbits close to a planet's rings and appears to keep the ring particles in place

shooting star A common name for a meteor

SMC The Small Magellanic Cloud

solar Relating to the Sun

solar eclipse An eclipse of the Sun in which the Moon partly (partial eclipse) or completely (total eclipse) covers the surface of the Sun and blots out its light; *See also* **eclipse**

solar flare A powerful explosion near the surface of the Sun that injects high-speed particles into the solar wind

solar system The Sun and the bodies that travel with it through space, primarily the planets, their moons and the asteroids

solar wind A stream of electrified particles given off by the Sun

spectral class The group in which a star is classified according to its spectrum

spectral lines Dark lines in the spectrum of a star or galaxy

spectroscope/spectrograph An instrument that splits light into a spectrum of different wavelengths, displaying the spectral lines

spectrum A band of rainbow colors produced when sunlight or starlight is split into separate wavelengths

speed of light The speed at which light and other electromagnetic radiation travels —around 186,000 miles (300,000 km) per second; it is the fastest speed possible

spiral galaxy A galaxy with curved arms coming out of a central bulge, or nucleus

standard candle An object that astronomers can use to estimate the distance to remote galaxies

star A huge globe of searing hot gases that gives off energy as light, heat and other radiation

stellar Relating to a star or stars

subatomic particle A particle that is smaller than an atom

sunspot A dark patch on the Sun's surface that is cooler than surrounding areas. The number visible vary according to an 11-year cycle, called the solar (or sunspot) cycle

supercluster A grouping of many clusters of galaxies occupying vast regions of space

supergiant The largest type of star, typically hundreds of times larger than the Sun

supernova A catastrophic explosion of a supergiant star (Type II supernova) or—even more powerful—of a white dwarf (Type I supernova)

terrestrial Relating to the Earth

terrestrial planets The planets Earth, Mercury, Venus and Mars, which are made up mainly of rock

trillion One million million: 1,000,000,000,000

true brightness *See* **absolute magnitude**

ultraviolet rays Electromagnetic rays that have a wavelength just shorter than that of visible violet light

universe Space and everything it contains: galaxies, stars, planets and energy

van Allen belts Donut-shaped regions of intense radiation around the Earth. There are similar regions on other planets

variable star A star that varies in brightness, usually because of an internal process; *see also* **Cepheid, eclipsing binary, Mira variable**

voids Vast regions of empty space between the superclusters of galaxies that make up the universe

wavelength The distance between the peaks or troughs of a wave motion, such as a wave of electromagnetic radiation

weightlessness The strange condition astronauts experience when traveling in orbit, in which their bodies appear not to have any weight, and gravity seems not to exist. This state is properly termed microgravity

white dwarf A small, very dense star, about the same size as the Earth but with a mass similar to that of the Sun; it marks a late stage in the life of Sun-like star

WIMP Weakly Interactive Massive Particle, thought to be one kind of dark matter

X-rays Penetrating electromagnetic rays with a wavelength longer than gamma rays but shorter than ultraviolet rays

zodiac An imaginary band in the heavens through which the Sun, Moon and planets appear to travel. Hence, the constellations of the zodiac

Landmarks in Astronomy

585 BC
Thales of Miletus correctly predicts the total solar eclipse of this year

350 BC
Aristotle proves that, since the Earth throws a curved shadow on the Moon, it is round and not flat

280 BC
Aristarchus of Samos suggests that the Earth might orbit the Sun, instead of vice versa

240 BC
Eratosthenes measures the size of the Earth with considerable accuracy

The first recorded appearance of what becomes later known as Halley's Comet

140 BC
Hipparchus creates a catalogue of over 1,000 stars and establishes the magnitude system of measuring the brightness of stars, which is still used today

AD 150
Ptolemy of Alexandria compiles the *Almagest*, which sums up ancient astronomical knowledge

813
Al-Mamun founds the Baghdad school of astronomy and translates the *Almagest* into Arabic

1054
Chinese astronomers witness a supernova in Taurus, which gives rise to the supernova remnant we know as the Crab Nebula

1433
Ulugh Beigh establishes the world's finest observatory at Samarqand (in modern-day Uzbekistan)

1543
Nicolaus Copernicus resurrects the concept of a solar system in his seminal work "De Revolutionibus Orbium Coelestium" (Concerning the Revolution of Celestial Spheres)

1572
Tycho Brahe observes a supernova in Cassiopeia

1595
David Fabricius discovers the variability of Mira Ceti

1600
Giordano Bruno is burned at the stake for the heresy of supporting Copernicus's concept of a solar system

1609
Johannes Kepler publishes the first of his laws of planetary motion, which states that planets move in elliptical orbits around the Sun

Galileo turns his telescope to the night sky and makes the first systematic survey

1666
Isaac Newton formulates his laws of gravity and investigates the visible spectrum

1672
Newton builds the first reflecting telescope

1675
Greenwich Observatory is founded in England

Ole Romer measures the speed of light

1682
Edmond Halley observes the bright comet that is subsequently named after him; he calculates that it is a regular visitor to the Earth's skies and (correctly) predicts its return in 1758

1728
James Bradley introduces the idea of using parallax to measure stellar distances

1781
William Herschel discovers Uranus, the first new planet discovered since ancient times

Charles Messier draws up a list of clusters and nebulas, to distinguish them from comets

1801
Guiseppe Piazzi discovers Ceres — the first and biggest asteroid

1802
William Wollaston observes dark lines in the Sun's spectrum

1815
Joseph Fraunhofer studies in detail the dark lines in the solar spectrum, henceforth known as Fraunhofer lines

1838
Using the parallax principle, Friedrich Bessel makes the first accurate measurement of stellar distance — 65 trillion miles, (105,000,000,000,000 km) to the star 61 Cygni

1843
Heinrich Schwabe discovers the 11-year solar (sunspot) cycle

1845
Lord Rosse completes his giant reflector ("the Leviathan of Parsonstown") at Birr Castle and discovers the spiral nature of some nebulas, including the Whirlpool Galaxy (M51)

1846
Johann Galle discovers an eighth planet — Neptune

1877
Giovanni Schiaparelli reports seeing "canali" (channels) on Mars, which is mistranslated from Italian as "canals," or artificial waterways; this gives rise to the idea of a Martian civilization

1888
Johann Dreyer's *New General Catalog of Nebulae and Clusters* (NGC) is published, which assigns numbers (that are still used) for the identification of clusters, nebulas and galaxies

1894
Percival Lowell founds the Lowell Observatory in Flagstaff, Arizona, specifically to study Mars

1905
Mount Wilson Observatory is founded in Pasadena, California

Ejnar Hertzsprung publishes the first of several papers about classifying stars

1908
The great Siberian "meteorite" lands; it is now thought to have been the nucleus of a comet

1912
Henrietta Leavitt discovers the period-luminosity law for Cepheid variables: that the period of variability of a Cepheid is related to its luminosity, or true brightness

1913
Henry Norris Russell publishes his ideas on stellar evolution, his work dovetailing with that of Hertzsprung; this leads to the famous Hertzsprung-Russell (H-R) diagram, which correlates stellar luminosity with spectral class or temperature

1917
The 100-inch (2.5-m) Hooker Telescope is completed at Mount Wilson Observatory

1918
Harlow Shapley makes the first accurate estimate of the size of the Milky Way Galaxy

1919
Edwin Hubble begins observing with the Hooker telescope at Mount Wilson Observatory

1920
Vesto Slipher announces that most galaxies have red shifts, indicating that they are all rushing away from us

1923
Hubble discovers a Cepheid in the Great Nebula in Andromeda and establishes that it is a separate star system — an external galaxy

1929
Hubble connects the red shifts of galaxies to their speed of recession and suggests that the universe is expanding

1930
Clyde Tombaugh discovers Pluto, which is classified as a planet until August 2006

1931
Karl Jansky detects radio waves coming from the heavens, which leads to the foundation of radio astronomy

1948
The 200-inch (5-m) Hale Telescope is completed at Mount Palomar Observatory, near Los Angeles

Hermann Bondi and Thomas Gold put forward the steady-state theory for the origin and evolution of the universe, later developed by Fred Hoyle. The theory states that the universe has always existed in the state it is now and will exist for ever in that steady state

1955
The first giant, steerable radio telescope (250 feet [82 m] in diameter) is completed at Jodrell Bank Observatory, near Manchester, England

1957
Russia's satellite Sputnik 1 is launched on October 4, pioneering the Space Age

1958
The United States launches its first satellite, Explorer 1, on January 31; the satellite discovers the van Allen belts of radiation circling the Earth

1959
Russia's Luna 2 probe crashes on the Moon; Luna 3 returns the first pictures of the lunar far side, always hidden from the Earth

1961
Russian cosmonaut Yuri Gagarin becomes the first human to orbit the Earth on April 12, which he does once, in a Vostok capsule

1962
John Glenn becomes the first US astronaut in orbit, circling Earth three times on February 21 in the Mercury capsule "Friendship 7"

The United States launches the first successful deep space probe, Mariner 2, which makes a distant but effective flyby of Venus

1963
Maarten Schmidt first identifies a quasar, while studying radio source 3C-273; he identifies it with a starlike object that has an astonishingly large red shift, proving that it is very much father away than a star can be

1965
Mariner 4 probe sends back the first close-up photographs of a planet — Mars

1967
Cambridge radio astronomer Jocelyn Bell-Burnell discovers the first pulsar

1969 Apollo 11 astronauts Neil Armstrong and Edwin ("Buzz") Aldrin become the first humans to set foot on the Moon

1973
Launch of the US space station Skylab; three crews there carry out an in-depth study of the Sun

1974 Pioneer 10 probe observes Jupiter from close quarters

1976
Two Viking probes reach Mars; orbiters survey the planet from orbit; landers touch down on the surface, take photographs, report on Martian weather and test soil (in vain) for signs of life

1977
Voyagers 1 and 2 are launched to explore the outer planets

1979
Pioneer 11 flies by Saturn, taking the first close-up pictures of the ringed planet

1981
The US space shuttle makes its debut on April 12, when orbiter Columbia thunders into space

1986
Voyager 2 reports from Uranus; Giotto (from Europe), Vega 1 and 2 (from Russia) and Sakigake and Suisei (from Japan) target Halley's Comet

1989
Voyager 2 makes its final planetary encounter, with Neptune

1990
Orbiter Discovery launches the HST on April 24, as part of shuttle mission STS-31

1993
Astronauts on shuttle mission STS-61 in December repair the HST in orbit, correcting its defective vision with COSTAR and fitting a new wide field/planetary camera and solar panels

1997
The second HST servicing mission (STS-82) takes place in February, involving the installation of a new spectrograph (STIS) and a near-infrared camera and spectrometer (NICMOS)

1999
Astronauts replace the HST's attitude-maintaining gyroscopes on the HST servicing mission STS-103 in December

2002
Astronauts install an advanced digital camera (ACS) to improve the quality of the HST's images on servicing mission STS-109 in March

2003
Shuttle orbiter Columbia breaks up just before landing on February 1, with the loss of seven crew. The three remaining shuttle orbiters are grounded and future missions put on hold

2008
The HST is to have a fifth and final servicing mission. Two new instruments will be installed and running repairs made

2013
The HST is expected to be at the end of its life; the next generation space telescope, the James Webb Space Telescope (JWST) is launched to replace it

Index

Acknowledgments

The author would like to extend his grateful thanks to the many individuals and organizations who have provided such invaluable assistance in the preparation of this book, particularly in the provision of images. Special mention must be made of Cheryl Gundy and the Public Outreach team at the Space Telescope Science Institute in Baltimore; Gwen Pitman and the photo research department at NASA headquarters in Washington D.C.; Mike Gentry and Gloria Sanchez of the Media Resource Center at the Johnson Space Center in Houston; Stephane Corvaja and Nadia Imbert-Vier at the European Space Agency in Paris; and Hans-Hermann Heyer of the European Southern Observatory's public relations department in Munich. Thank you all!

Picture Credits

(t=top, b=bottom, l=left, r=right, c=center)

All Hubble Space Telescope images are credited to NASA and the following additional teams and institutions.

ACS is an abbreviation for the Advanced Camera for Surveys team
AURA, the Associated Universities for Research in Astronomy
COBE, the Cosmic Background Explorer satellite
CXC, Chandra X-Ray Center
ESO, the European Southern Observatory
ESA, the European Space Agency
GFSC, the Goddard Spaceflight Center
GUGEG, the Göttingen University Galaxy Evolution Group, Germany
HH, the Hubble Heritage team (STScI/AURA)
HTOTPT, the Hubble Telescope Orion Treasury Project Team
IAA, the Instituto de Astrofísica de Andalucía, Spain
JAXA, Japan Aerospace Exploration Agency
JPL, the Jet Propulsion Laboratory
JSC, the Johnson Space Center
KIPAC, the Kavli Institute for Particle Astrophysics and Cosmology
NASA, the National Aeronautics and Space Administration
NRAO, the National Radio Astronomy Observatory
SETI, Search for Extra-Terrestrial Intelligence
SP, Spacecharts Photolibrary
SSI, Space Science Institue
STScI, the Space Telescope Science Institute
U., University

Page 2 STScI, La Plata Observatory; 4 ESA, HH; 6 HH; 12 HH; 14 HH; 16 ESA, Arizona State U.; 17 ESA, Arizona State U.; 19 ESA and HH; 20 ACS, ESA; 21 ESA, HH; 22 HH; 24tr ESA, STScI and HTOTPT; 24b Harvard-Smithsonian Center, Steward Observatory, Rice U.; 25 ESA, STScI and HTOTPT; 26t HH; 26b ESA and HH; 27r HH; 29t STScI; 29b HH; 29c U. of Colorado; 29b Caltech, JPL; 30, 31 ESA and HH.; 32 STScI; 33 ESA and GUGEG; 34t JPL-Caltech, STScI, Rochester Inst. of Technology, U. of Wisconsin and the GLIMPSE Legacy Team; 34b John Hopkins U.; 35 ESA, U. of British Columbia; 36t ESA, IAA; 36b STScI; 37 ESA, STScI; 38 HH; 39tb STScI; 40 ESA, STScI; 41: ESA, STScI, Catholic U. of Leuven (Belgium), U. of California (Berkeley;) 43 U. of Illinois, U. of Washington; 44 U. of Arizona, NICMOS team; 45b Rice U.; 46 HH; 47 HH; 48 ESA, HEIC and HH; 49l STScI; 49r IHH, Nordic Optical Telescope; 50 HH; 52 JPL; 53 ESA and Albert Zijlstra; 55 ESA, Vanderbilt U.; 57 ESA and HH;58 U. of Colorado; 60t: ESA and HH; 60b STScI; 61 ESA, HH; 62 ESA, Harvard-Smithsonian Center for Astrophysics; 63r Arizona State U.; 64, 65 ESA and HH; 67 ESA; 69r STScI; 70 HH; 71l ESA; 71r STScI;72 U. of Maryland, Caltech, Sabaru Telescope, U. of Leicester; 73t HH; 73b Tel-Aviv U., Columbia U.; 74 ESA, GUGEG; 76c ESA and MPIA; 77bl STScI; 77br ESA, ESO; 78t NRAO; 78b STScI; 79l Princeton U., U. of Wales, STScI; 79r STScI; 80t U. of Alabama, Gemini Observatory, NRAO; 80b Yale U.; 81 HH; 82b STScI; 84 HH, ESA; 85b ESA, HH; 86 ACS, ESA; 89 ESA, HH; 90 ESA, Tel Aviv U., Caltech; 91 H. Richer (U. of British Columbia); 92 K. Lanzetta; 93 ESA, STScI, the HUDF Team; 94tl JPL; 94 tr Hinode JAXA, PPARC; 95 ESA, John Hopkins U.; 97 STScI; 98 STScI; 99 STScI; 101 ESA, HH; 102t GFSC, STScI; 102b Carnegie Observatories, Columbia U.; 104 ESA, HH; 105l STScI; 105r U. of Manitoba, Pennsylvania State U., U. of Arizona, STScI; 106t STScI; 106b ACS, Hebrew U., John Hopkins U., Lick Observatory; 107 U. of Hawaii, STScI; 108t Caltech, Observatoire Midi-Pyrenees, ESA; 109 CXC; 110 ESA, HH; 112 GFSC, STScI;113 ESA, STScI; 114 STScI; 115t ESA, U. of California (Berkeley); 115b NASA, ESA, ACS Science Team; 117t Caltech, JPL; 117t ESA, HH, Cornell U., SSI; t117b ESA, U. of Arizona, STScI; 122t Applied Research Corp., STScI; 123tl JPL-Caltech; 123tr Rudi Vavra; 123b ESA, JHU, APL, STScI; 124, 125 HST Comet Science Team; 126 HH; 127l JPL; 127cb STScI; 130 Cornell U., U. of Arizona; 136t ESA, HH; 136b U. of Colorado, U. of Toledo, STScI; 137br ESA, DLR, FU Berlin; 137tl JPL-Caltech, Cornell, NMMNH; 138 Cornell U., HH; 139t ESA, U. of Arizona; 139bl ESA, U. of California, Berkeley; 140 HH; 141t JPL; 141c JPL, U. of Arizona; 142t U. of Arizona; 142b Cornell U., U. of Arizona; 143t U. of Arizona; 143b JPL; 144t ESA, U. of Wisconsin, SSI, SETI Institute; 145b ESA, SETI Institute; 146l and r ESA, U. of Arizona, SSI; 147tc ESA, JHU, SwRI and the HST Pluto Companion Search Team; 147tr ESA, Caltech 147cl ESA, STScI;151 ESA, Space Telescope European Coordinating Facility, Germany; 153bl JPL-Caltech, Steward Observatory; 153br: JPL-Caltech 154: x-ray: CXC, KIPAC; radio: NRAO, VLA; Infrared: ESA, McMaster U; 156lc JPL-Caltech, U. of Arizona; 156b COBE Project, DMR; 157b JAXA; 158tl ESA; 158b JPL-Cornell; 159tl JPL-Caltech, UMD 159tr and bl JPL, U. of Ariz; 161bl JPL,SSI; 161cl JPL SSI; 168b STScI; 169 HH; 170t HH; 170b U. of Colorado, U. of Toledo; 171t ACS; 171b STScI; 172br STScI; 176t STScI; 177 UCLA; 179l ESA and STScI; 180 ESA, McMaster U.

Additional images are credited to the following sources:

9 JSC; 18 SP; 27l ESO; 45t SP; 54 SP; 59 SP; 63l SP; 66 SP; 68l NRAO; 69l ESO; 76 SP; 82t NRAO; 82r ESO; 88bl SP; 88br NRAO; 96b SP; 100 SP; 103 SP; 108b SP; 116 SP; 120, 121 SP; 122b Martin Moberley; 127tr SP; 128, 129 SP; 132, 133 SP; 134, 135 SP; 144b SP; 143tc SP; 147b SP; 148 ESO; 150 W.M. Keck Observatory; 152 Robin Kerrod; 152t NRAO;153tl NRAO, AUI; 154 SP; 155 ESAC and ESA; 157t Robin Kerrod; 163 JSC; 164 SP; 165 JSC; 167 SP; 168t SP; 172t JSC; 172bl JSC; 173 JSC; 174, 175 JSC; 176b JSC; 179r JSC